通往文人的建造

传统与当代的设计思辨

潘冉 · 著

东南大学出版社

The
Construction
Leads
To
Humanity

Taditional and
contemporary
design
speculations

Pan Ran

Southeast University Press

我理解

世上最难舍的希望

是太阳从阴霾处透出的第一缕光

就像

热爱着地球的建筑师

被自然原谅

和林木和解

在美好的事物面前

保持克制

与冷静

五年前，明媚的下午，树影婆婆。我坐在那个结满青果，并散发着阵阵幽香的苦楝树下，眼前的院子像是一个绕过苏醒的世界，青砖条石，越过百年，据说这是那年整条街上唯一一个教书先生的宅邸。静谧于建造者，更能连接未知，不自觉地提起笔来，抓过一张白纸，涂抹起稍纵即逝的样子。"来"，动词，由远到近，预示着从传统到当代，"来院"在思索的叙事中诞生了。很显然，我是幸运的，之后的五年，只是日常地刻画叙事的受光面。从几何空间与石板青砖构成的恢宏暗喻，到青竹浓郁光影斑斓的润物无声，每一天，被空间眷顾并滋养着，与枝叶一起生长，朝阳晨露，秋雨泥土。水痕是自然的皴法，盛夏的虫子归于月光，而我们在"来院"的空间里，日复一日地厮磨着建造的点滴，与工匠，与友人，与伙伴。人归根结底是骄傲的，企图通过生命运动的轨迹证明自己存在于时代的痕迹。于是，建造者将建筑放平，音乐家将音符挂起来……

Five years ago, on a bright afternoon, the trees were whirling, and I sat under the neem tree full of green fruit and fragrance. The yard in front of me was just like a sleeping world with green brick and stone, Over hundreds of years, the yard is said to be the only mansion of a teacher in the whole street in that year. The silence, as for the builder, is better to connect the unknown. I unconsciously picked up the pen, grabbed a blank piece of paper, and painted the picture with a fleeting look. "Lai", a verb, which in chinese means from far to near, indicates the tendency from the traditional to the contemporary. "Lai Yard" was born in the contemplative narrative. Obviously, I was lucky. For the next five years, I only depicted the bright side of the narrative on a daily basis. Water marks are a natural texturing method, and the bugs in midsummer belong to the moonlight. We have been working together for the construction details day after day with craftsmen, friends and partners in "Lai Yard". In the final analysis, people are proud. They try to prove their existence in the traces of the times through the track of the movement of life. Therefore, the builders flattened the buildings and the musicians hung up the notes...

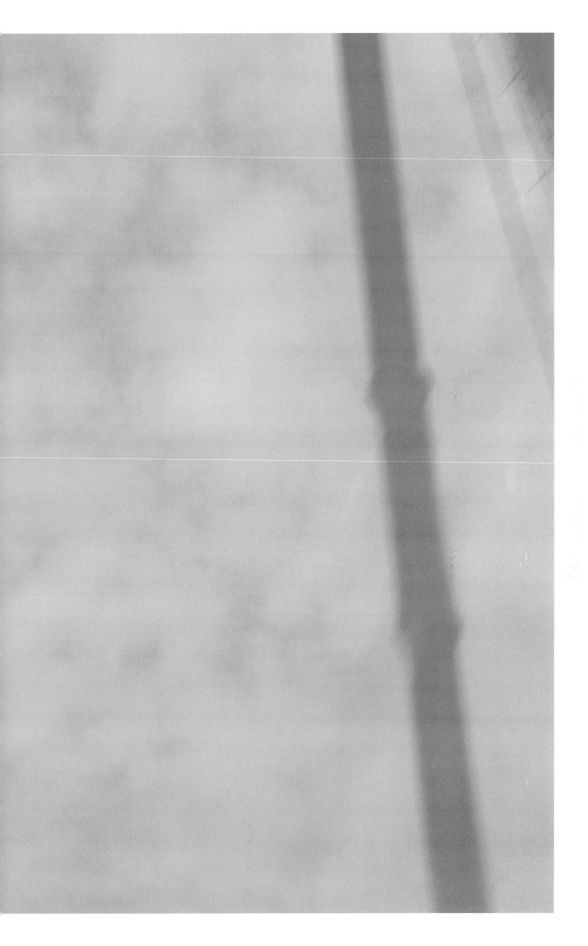

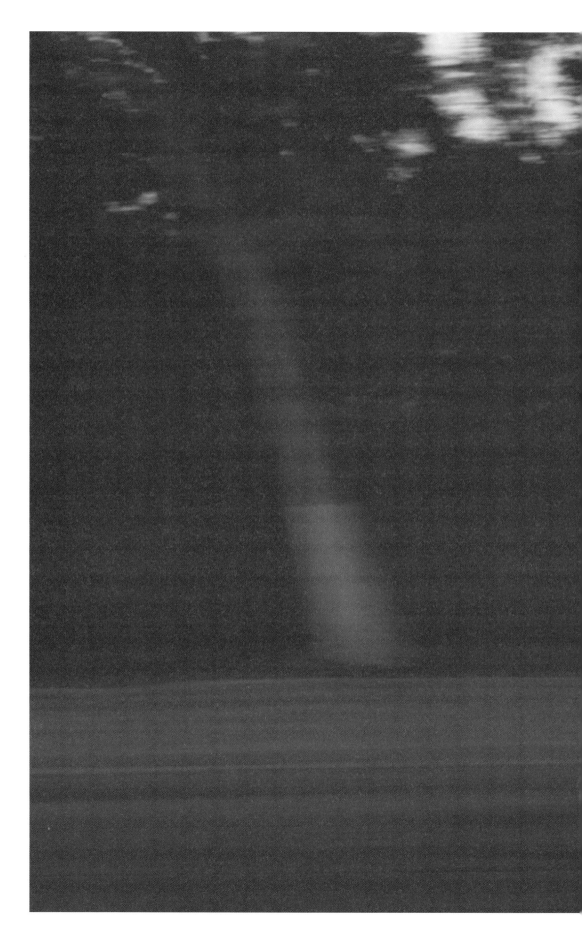

0 2 4 通过设计究竟可以做到什么?

0 2 6 What can be accomplished through designing？

0 3 3 大江中岳·及道
JI Dao

0 6 9 光和空的意志·来院
LAI Yard

0 9 1 竹里馆
Bamboo's Eatery

1 1 5 印象村野
Village Impression Restaurant

1 3 5 长物之宅
Residence of Zhang Wu

1 7 7 全息的觉知·二十四单
24 Cathay Restaurant

2 0 3 红公馆造店记之烟雨江南
Misty Rain of JiangNan Made by Hong Mansion

2 3 9 红公馆造店记之胭脂水粉
Rough Gouache Made by the Hong Mansion

2 7 5 重塑构成美学 My Hotel 的触觉札记
Reshaping the Aesthetics of Composition — Tactile Notes of My Hotel

2 9 9 光之叙事
Story of Light

3 1 5 时间无垠 予筑新生 —— 四十号
#40

3 5 6 建筑不是妥协, 设计更需要克制

3 6 4 静谧与光明的交响 —— 路易斯·康建筑之旅

3 6 5 Symphony of Silence and Light — Journey to Louis Kahn's Architecture

3 6 6 耶鲁大学美术馆 —— 古典与现代的完美结合

3 6 8 Yale University Art Gallery — Perfect Fusion of the Classic and Modern

3 7 0 耶鲁大学英国艺术中心 —— 不朽的遗世之作

3 7 2 Yale Center for British Art — Monumental Masterpiece

3 7 4 理查德医学研究中心 —— 竖向秩序

3 7 5 Richards Medical Research Laboratories — Vertical Order

3 7 6 索尔克生物研究所 —— 混凝土与砾石的神庙

3 7 8 Salk Institute for Biological Studies — Temple of Concrete and Gravel

3 8 1 亚平宁建筑之旅 13 日 —— 关于卡罗·斯卡帕

3 8 2 A 13-day Architectural Tour in Apennine — About Carlo Scarpa

3 8 4 Castelvecchio 城堡博物馆 —— 半遮面的遗憾

3 8 5 Castelvecchio Museum — Half-concealed Beauty with Imperfection

3 8 6 Brion 墓园 —— 生与死的终极浪漫

3 8 8 Brion Cemetery — The Ultimate Romance of Life and Death

3 9 0 Stampalia 基金会 —— 叠加后的艺术馆

3 9 2 Stampalia Foundation — A Museum of Superimposition

3 9 4 番外 —— 威尼斯的商业街道

3 9 5 Another Story — Commercial Streets in Venice

3 9 6 设计感悟之建筑与时装

3 9 6 Sentiment of Design:the Architecture and the Clothes

3 9 8 设计漫谈

4 0 0 Thoughts on Designing

4 0 2 设计圈的学习之道 —— 关于读书和旅行

4 0 6 Ways of Study in Design Community — About Reading and Travelling

4 1 0 现代装饰 2014 年 5 月刊·对话　来自上海读者的提问

4 1 1 *Modern Design*　May, 2014 Dialogue column
Questions from the Readers　from Shanghai

4 1 2 聚焦江南 回归东方精神本源

通过设计
究竟可以做到什么？

怀旧更多的不是回到某种熟悉的过程，
而是进入与过去有所联系的未来……

日光下行走，云卷时思考，现实中造梦。设计为社会的沟通而努力。通过设计究竟可以做到什么？一个长期存在于脑海中的问询。我们创造空间，代表架构新的价值。笔触代言着城市的欲望，某种程度上反映出社会主导价值取向的素养。"设计"不应被理解为孤立的审美行为，专属表达自我的欲望和快感，比起做东西，更多工作应该侧重于解析"环境"和"条件"。思考目标不局限于建造一个功能空间，它可以是一个想象的场所，具有足够的包容性，引导人和环境共同成长。职业本能引导我思考怎样才能创造出不断自我更新的形象，留下恒久印象的清澈通透的痕迹，同时我的所有作品都反映了既定环境的属性。大多数人会下意识地认为美常驻足于创造行为的领域，但我始终相信，优雅之美绝非短期努力所能达到的成就，其创造只能通过漫长的打磨过程。正如钟乳石洞是由一滴一滴下落的水珠重复累积才形成的，环境空间的精彩亦是变化着的世界的精神影响的逐渐累积。

我们生活在城市文化的包裹下，城市文化最重要的意义在于它作为人们的原始记忆，超越时代流传下去。土地存在记忆，"地缘"便成了我们做设计无法回避的要素。设计师是世界的，设计则有其区域性。金字塔在埃及，泰姬陵在印度，长城属于中国，这些都绝非偶然。新设计成果的成立与否，取决于创造者对原有地域文化的尊重态度。要在一块土地上营造新的事物，需采取某种方式来对应这块土地原本的多元价值观所积累的"场域记忆"。新与旧之间的对话，能让场域活性化，随之产生更强的空间深度。诚然，保留旧事物并使之能够在现代重生，绝对比新的建设来得耗时费工，但人是凭借着小小的回忆活着的，为了精神上的满足，难道不应努力恪守职责，留下记忆的痕迹么？

恪守不变是否就是怀旧？不，怀旧更多的不是回到某种熟悉的过程，而是进入与过去有所联系的未来。历史推进，社会进步，人类生活方式发展，就要求对已有的设计思维进行不停歇地拓展。区别于城市空间，建筑空间和室内空间则是更敏感的接触使用者行为习惯的第一层表皮。一个空间艺术的营造，建成或只是起点，使用的过程才是真正展现出其艺术魅力的重点。满足使用者的要求，必须是凌驾于自我实现的第一出发点。使用方式与营造空间发生冲突时，创造的环境必定凋零；但当使用方式与营造空间契合默契时，就相当于给空间注入了灵魂，各种元素都蕴藏着发展的可能性。设计不可维持被动姿态，需主动唤醒人们心中对生命的关爱和周遭环境的意识，在塑造环境时，最后的依靠往往就是人们对环境空间的感受力。设计师，为社会的沟通而努力，是否至少将我们的"田地"收拾好？

What
can be accomplished through
designing ?

Walk under the sun, ponder on the clouds, and create when back on earth. Design serves for the communication of the society. What can be accomplished through designing? It's a question I've long been thinking about. We create space, represent and construct new values. Our designs speak for the desires of cities and to some extent reflect the quality of the society's leading values. "Design" should not be understood as an isolated aesthetic act. It does not serve for mere expressions of personal longing and sensation. Instead of focusing on the process of design, we should pay more attention to analyze the "context" and "condition" of it. We should not limit our thinking to only construct a functional space. It can be an imaginative place, inclusive enough to lead the common progress of both the human being and the environment. My work ethic leads me to reflect on how to create self-renewable identities, to retain everlasting and crystalline traces, and all my works reflect the nature of a given environment. Most people subconsciously think that beauty belongs only to the creation process, but I hold the belief that beauty is not a product of short-term efforts. Instead, it is created over a long period of polishing and refinement. Just as the stalactite is formed by constant deposition of the dripping water, the fascinating environmental space is shaped by the constant spiritual influence of the changing world.

We live within the city culture, whose most important value lies in its potential to be passed down throughout generations as an original memory of human beings. The land reserves the memory, so "the geography" becomes an indispensable part of our design. The designers belong to the world but the design is shaped by the region. The Pyramid belongs to Egypt, Taj Mahal belongs to India and the Great Wall belongs to China. These architectures do not just happen to be where they are now. The success of a new design comes from the respect that the designers pay to the local culture. When setting up a new project on a piece of land, we need to adopt a

We are not nostalgic for the past happenings;
rather, we are trying to find a prospect for the
future which is linked to the past.

certain means to deal with the "spatial memory" shaped by the diverse local values. The conversation between the old and the new will revitalize the space and generate deeper spatial consideration. It is, admittedly, more time-consuming to preserve the old things and try to revitalize them in modern times compared to simply constructing new ones. But it is with those trivial memories that we human beings are able to live a life. Therefore, shouldn't we make our utmost efforts to save the traces of memories for our spiritual fulfillment?

Does the loyalty to the past necessarily mean nostalgia? The answer is no. We are not nostalgic for the past happenings; rather, we are trying to find a prospect for the future which is linked to the past. The history moves on, the society progresses and people's way of life develops. All of these facts motivate the designers to adjust their design logic constantly. Different from the urban space, the architectural space and the interior space are the first layer of skin that touches upon the behaviors of the users in a more sensitive way. The creation of an artistic space marks only the beginning of the design journey. The artistic charm indeed lies in the functioning process. To meet the need of the user must be the priority for designing instead of to achieve the self-fulfillment of the designer. When the functioning part clashes with the construction process, the creation becomes unfeasible. But when the former matched the latter perfectly, a space becomes intelligent and all of its elements are endowed with possibilities for future development. In the design progress, we should not assume a passive attitude; rather, we should take the initiative to arouse people's care toward life and their awareness toward the surrounding environment. After all, our final reliance in constructing a space lies in people's ability to perceive the environment. The designer endeavors to facilitate the communication of the society. Therefore, shouldn't we at least tidy up our own mind first?

Starting from the traditional construction method to the internal relationship between the field of art and architecture, Mr. Pan Ran has a high degree of unity of a complete and clear cognitive design context and design practice. He is known as the representative of promotion and practice of the oriental spirit of contemporary architectural interior design.

Representative works: Lai Yard、 Xiao Dong Yuan、 Residence of ZhangWu. The humanistic atmosphere was released from his works, concealing the internal relationship between man and nature. His works also like to interpret space aesthetics and philosophy from the exquisite life attitude of Jiangnan literati.

潘冉先生从传统筑造功法到艺术领域与建筑的内在联系出发，完整清晰的认知设计学脉络与设计践行高度统一，被誉为当代建筑室内设计东方精神推动与践行之代表人物。

代表作：《来院》《小东园》《长物之宅》。作品中透出的人文气息暗藏人与自然的内在关系，并喜从江南文人墨客的精致生活态度中解读空间美学与哲思。

潘冉

Pan Ran

Five years ago, on a bright afternoon, the trees were whirling, and I sat under the neem tree full of green fruit and fragrance. The yard in front of me was just like a sleeping world with green brick and stone, Over hundreds of years. the yard is said to be the only mansion of a teacher in the whole street in that year. The silence, as for the builder, is better to connect the unknown. I unconsciously picked up the pen, grabbed a blank piece of paper, and painted the picture with a fleeting look. "Lai", a verb, which in Chinese means from far to near, indicates the tendency from the traditional to the contemporary. "Lai Yard" was born in the contemplative narrative. Obviously, I was lucky. For the next five years, I only depicted the bright side of the narrative on a daily basis. Water marks are a natural texturing method, and the bugs in midsummer belong to the moonlight. We have been working together for the construction details day after day with craftsmen, friends and partners in "Lai Yard". In the final analysis, people are proud. They try to prove their existence in the traces of the times through the track of the movement of life. Therefore, the builders flattened the buildings and the musicians hung up the notes...

五年前，明媚的下午，树影婆娑。我坐在那个结满青果，并散发着阵阵幽香的苦楝树下，眼前的院子像是一个绕过苏醒的世界，青砖条石，越过百年，据说这是那年整条街上唯一一个教书先生的宅邸。静谧于建造者，更能连接未知，不自觉地提起笔来，抓过一张白纸，涂抹起稍纵即逝的样子。"来"，动词，由远到近，预示着从传统到当代，"来院"在思索的叙事中诞生了。很显然，我是幸运的，之后的五年，只是日常地刻画叙事的受光面。从几何空间与石板青砖构成的恢宏暗喻，到青竹浓郁光影斑斓的润物无声，每一天，被空间眷顾并滋养着，与枝叶一起生长，朝阳晨露，秋雨泥土。水痕是自然的皴法，盛夏的虫子归于月光，而我们在"来院"的空间里，日复一日地厮磨着建造的点滴，与工匠，与友人，与伙伴。人归根结底是骄傲的，期望通过生命运动的轨迹证明自己存在于时代的痕迹。于是，建造者将建筑放平，音乐家将音符挂起来……

大江中岳 · 及道

项 目 名 称	及道艺术馆
陈设设计 & 执行	名谷设计机构蜜麒麟陈设组
特 约 艺 术 家	乐泉
项 目 地 点	南京老门东历史街区边 37-1 号
项 目 面 积	室内 120 平方米，庭院 80 平方米
主 要 材 料	稻草泥 实木板 钢板 章子纸
竣 工 日 期	2019.7
摄 影	夏至 全仲贤

JI Dao

Project name: JI Dao Museum of Art

Display design & execution: Minggu Design Agency, Miqilin
Display Group

Special guest artist: Le Quan

Project location: No.37-1, Laomendong Historical Street,
Nanjing, China.

Project area: Indoor 120m², courtyard 80m²

Primary materials: straw paste, solid plank, steel plate, zhangzi paper

Completion date: 2019.7

Photographer: Xia Zhi, Quan Zhongxian

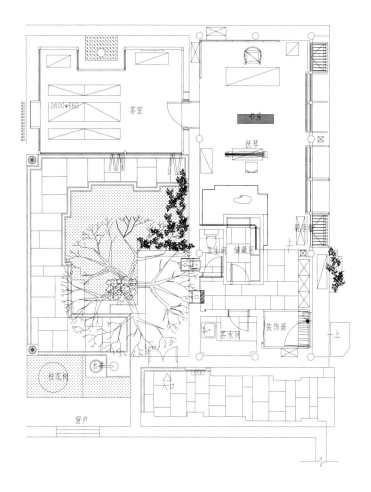

〔造亭子〕

    从边营 37-1 号的紫竹林进入，尽头墙面的砖刻"鸿禧"二字已被蕉叶遮蔽。往前一步，跨过右侧门廊，峰回路转见得开朗，原是个一眼看穿的长方院子。入口枇杷，墙角苦楝，屋檐下探，树冠层叠以蔽日，苍古又生机。踱步其中，三年一瞬，于树下建造一个"亭子"的欲望日渐滋长。一个有机空间将生活日常有效参与自然变化的载体，康说"建筑存在于建筑之前"，像大树生发出新枝，本应该存在的那个部分。建造的欲望源于限定，当建筑停止遮蔽，空间就会变得肆无忌惮，而亭子象征着自由者对自己的约束，一面显山露水，一面不越雷池。

    以"亭"的概念为延伸，新筑伏于树冠之下，依东墙而建，四周剥离开原建筑墙体，平面让出树的生长空间，围出一方天井，雨水可自然落入，地面抬高并退让出与立面相应的尺度关系，与南面新增一进院采用线性边界的连通方式，拒绝整面接触，最大限度地避免新筑对原建的破坏。顶面开放出与室内茶桌同等尺度的采光界面，并转折至北立面，秋冬两季树叶尽落，阳光可从顶面直达室内，空间获得更为直接的采光，并有效调节了室温，落座仰视可见枝干与天空叠合的流动之美。春夏季枝叶茂盛，烈日透过树荫的遮蔽，投向室内斑斓的影，大自然的涂抹在季节变化中，不断地给予触觉新知。

〔安于闹市〕

　　新增一进五十见方砖木结构建筑，北邻内院，与"亭"相连，宁静清幽，南临街巷，游人嘈杂、内外由花格门区隔，保留原建的梁架关系。于两侧柱跨的顶面安置设备，中间柱跨可完整获得古建的檩、椽结构。在西侧的五架梁下，中置建立一个盒体作为冥想空间，由西向东悬于地台之上，取临水小筑之意境，建立精神核心，空间随尺度的开合呼应，被无形地划分为东西两个区域，并由南向受光立面贯穿始终。南立面针对需要解决的多个功能问题，分别规划出视觉层、冷暖层、柔光层三个界面，并有机地叠合起来，满足自然通风的推窗藏于视觉层之后，柔光层可根据需要引入自然光线的起落变化，同时可避免外部干扰，保持视线界面的纯粹。盒体北侧连接墙体形成一个隐藏式空间，设为洗手间，通过凿壁采光的方式将北院光线转换进入过道的尽头，成为引导进入洗手区的唯一途径。南侧留出连接西南入口的室内交通，一个幽暗且神秘的入口以与古为新的方式完成由外部街巷的偶然随机性到隐于街市的克制平静，过渡到内院并获得避世般的自省空间。

　　小筑同样可以海纳百川，亦可气象深远，就像一段十秒钟的乐句，三秒钟的音符律动，便有可能获得持久的曼妙。对于母题"江南"的探讨是长期而持续的，她可以延绵不断地提供画面与直觉，让我轻松地告知自己，希望存在于怎样的环境氛围中，随后去建立与场景相关的建造逻辑。当在一个空间中置入另一个空间，从而一站式解决原空间中需要解决的一系列问题，比如分割空间、建立节奏、增加使用的机能、制定必然的动线、创造新的机会以隐藏现代设备等，如此以中国道家中"四两拨千斤"的方式解决建造的重复性，干净利落而直达目标。建筑态度亦是明确的，强调机能效率，骨骼意识高于表皮面貌，反对无限制地深入。

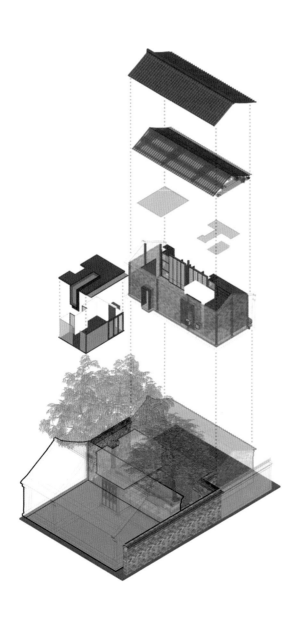

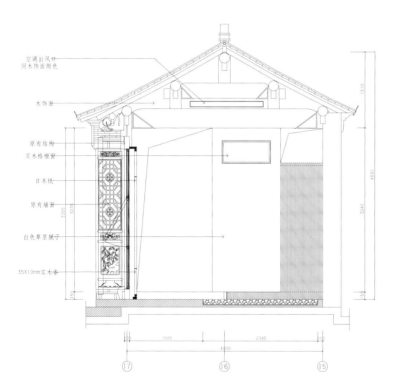

空调出风口
同木饰面颜色

木饰面

原有结构
实木格栅窗

日本纸

原有墙面

白色草茎腻子

35X10mm实木条

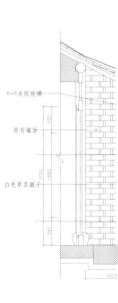

6+6夹胶玻璃

原有墙面

白色草茎腻子

① ① ①
17 16 15

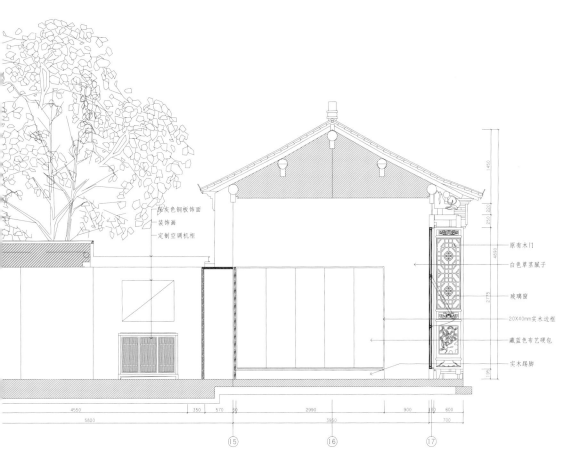

浅灰色钢板饰面
装饰画
定制空调机柜

原有木门
白色草茎腻子
玻璃窗
20X40mm实木边框
藏蓝色布艺硬包
实木踢脚

4550   350 570 50   2990   900 100 600
5820   3950   700

⑮   ⑯   ⑰

Build Pavilions

Entering from the purple bamboo forest at No. 37-1 of Bianying street, the words "Hongxi"engraved on the wall at the end have been obscured by banana leaves. One step forward, across the right porch, the peak turns to be bright and cheerful. It is originally a

long courtyard with the loquat at the entrance, with bead trees in the corner. Leaves probe under the leaves, and the canopy stackes to cover the sun, antiquity and vitality. Walking through it, three years has passed in a flash. The desire to build a pavilion under the tree grows day by day. An organic space is a carrier for daily life to effectively participate in natural changes. Kang said that "architecture exists before the building." It's like a big tree giving birth to new branches which are the parts that should have existed. The desire to build comes from limitation. When the building stops shading, the space will become unscrupulous. However, the pavilion symbolizes the restraint of the free man, showing off the highlights on one side while never going beyond the prescribed limit on another side.

With the concept of pavilion as an extension, the new building is built under the tree crown, relying on the east wall. The building peels off the original building wall surrounded and the

plane gives way to the growth space of the tree, encircling a patio which rain water can fall in naturally. The ground is raised and retreated to give scale to the facades correspondingly. The connection with the new yards in south under the mode of linear boundary refuses the contact of the whole side, and avoids the damage of the new building to the original building

院中的枇杷树长势喜人

The loquat trees in the courtyard are flourishing.

to the maximum extent. The top surface opens the daylighting interface of the same scale as the indoor tea table, and turns to the north facade. The leaves fall in autumn and winter, and then the sun can reach the room directly from the top. Thus, the space get more direct daylighting, and effectively adjust the room temperature. Sitting down and looking up, you can see the beauty from the flow of branches overlapping with the sky. The branches and leaves are luxuriant in spring and summer, and the scorching sun casts into the beautiful shadow of the room through the shade of the trees. The smear of nature constantly gives new feelings to touch in the changing seasons.

Settle in downtown

A new brick and wood building is added, 50m², which is adjacent to the inner courtyard in the north, and connected to the pavilion, quiet and peaceful. The building is located besides streets and alleys in the south where the noisy visitors pass by. The inside and outside space is separated by checkered doors, retaining the original beam-frame relationship, and equipment is placed on the top of the columns on both sides. The middle column span can completely obtain the purlin and rafter structure of the ancient building. Under the five beams on the west side, a box is built in the middle as a meditation space, suspended from west to east on the platform. The artistic conception of a small building near the water is taken to establish a spiritual core. With the opening and closing of the scale, the space is invisibly divided into east and west areas, and from the south to the light facade runs through all the time. In view of the multiple functional problems that need to be solved, the south elevation creates three interfaces: visual layer, cold and warm layer and soft light layer, which are stacked organically. The push window that meets the natural ventilation is hidden behind the visual layer. The soft light layer can bring the rise and fall of natural light according to the needs, while avoiding external interference and keeping the purity of line vision interface.

亭子象征着自由者对自己的约束，一面显山露水，一面不越雷池

The pavilion symbolizes the self-restraint of free individuals, who show themselves on one side while never crossing the line on the other.

入口枇杷, 墙角苦楝, 屋檐下探, 树冠层叠以蔽日, 苍古又生机

The loquat at the entrance, the chinaberry at
the corner, the downward facing leaves, the
stacked canopy of trees that shelter the sun, all
look old, yet vigorous.

康说"建筑存在于建筑之前"，像大树生发出
新枝，本应该存在的那个部分。建造的欲望
源于限定，当建筑停止遮蔽，空间就会变得
肆无忌惮

Kang said that "architecture exists before
architecture", like the part that should be
there before a tree produces new branches.
The desire to build arises from limits, and
when a building is no longer covered, space
becomes unbridled.

顶面开放出与室内茶桌同等尺度的采光界面，并转折至北立面。秋冬两季树叶尽落，阳光可从顶面直达室内，空间获得更为直接的采光，并有效调节了室温、落座仰视可见枝干与天空叠合的流动之美。春夏季枝叶茂盛，烈日透过树荫的遮蔽，投向室内斑斓的影，大自然的涂抹在季节变化中，不断地给予触觉新知

The roof opens up to the same size as the interior tea table for light, which extends and turns to the north facade. In autumn and winter when leaves fall, the sunlight can reach the room from the top, providing more direct lighting and regulating the room temperature. In the spring and summer, the foliage is lush and the sun shines through. The smear of nature constantly gives new feelings to touch in the changing seasons.

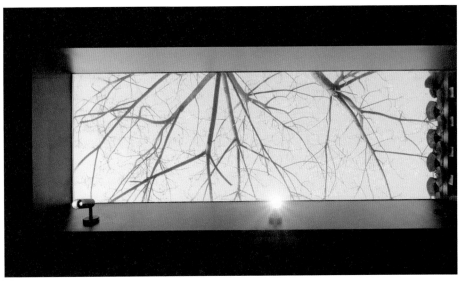

内外由花格门区隔，保留原建的梁架关系，于两侧柱跨的顶面安置设备，中间柱跨可完整获得古建的檩、椽结构

The interior and exterior are separated by lattice doors, retaining the original structural beams, with equipment placed on top of the columns on both sides; the middle pillar can completely adopt the ancient purlin and rafter structure.

南立面针对需要解决的多个功能问题，分别规划出视觉层、冷暖层、柔光层三个界面，并有机地叠合起来，满足自然通风的推窗藏于视觉层之后，柔光层可根据需要引入自然光线的起落变化，同时回避外部干扰，保持视线界面的纯粹

To address a number of functional issues, three interfaces, namely the visual layer, the cooling and warming layer, and the soft light layer, are planned and organically integrated in the south facade; the sliding window for natural ventilation is hidden behind the visual layer, and the soft light layer can introduce changing natural light as necessary, while avoiding external interference to keep the visual layer purity.

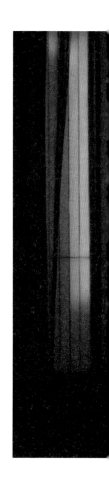

在西侧的五架梁下，中置建立一个盒体作为冥想空间，由西向东悬于
地台之上，取临水小筑之意境，建立精神核心。空间随尺度的开合呼
应被无形地划分为东西两个区域，并由南向受光立面贯穿始终

Under the five beams on the west side, a box is built in the middle for
meditation, suspended from west to east on the platform, resembling the scene
of a small building near the water and establishing a spiritual core; the space is
invisibly divided into two regions, east and west, with opening and narrowing
changes based on the size, both extending towards the light receiving facade in
the south.

凿壁采光

Chisel into a wall for lighting

一个幽暗且神秘的入口，以与古为新的方式完成由外部街巷的偶然随机性
到隐于街市的克制平静

A dark and mysterious entrance delivers the randomness of the outside
streets to the restrained calm of seclusion in a new way based on the old.

骨骼意识高于表皮面貌，反对无限制地深入

Skeletal consciousness prevails over superficial appearance and opposes unrestricted penetration.

项目地点	南京市老门东历史街区
项目面积	1 000 平方米
竣工日期	2015.7
主要材料	钢板 砖瓦石 泥灰 木板

LAI Yard

Project location: Laomendong Historical Street, Nanjing City

Project area: 1 000m²

Completion date : 2015.7

Primary materials: steel plates, brick stone, plaster, plank

光和空的意志 · 来院

位于城南中营的朴素古宅，与热闹的名号迥异，其实性格内向。与古城墙为邻百年，默然仁立巷口，于风雨飘摇之际被列为保护建筑得以修缮，北侧加建两栋仿古建筑，共组三进式院落，入口古朴，尺度窄小，通过时低头，抬头时开朗，院内树木建筑交织映衬，和谐优雅。随机缘为名谷设计机构进驻。客观来说，仿古建造的第二进"来院"建筑基底并非优越。工艺的精准度、材料的运用不及古人的手工制作，加之缺少时间的冲刷洗练，与真迹并肩多少夹杂一丝尴尬。即便如此，它仍反映了当下这个时间空间内人们对传统质朴的追念与渴望。

Located in the south of city, the plain historic house is introverted, which is different from its lively name in fact. Hundreds of years, silently standing at the alley and near the ancient city wall, it was listed as the protection buildings to be repaired. On the north of the original house, two pseudo-classic architectures were built to constitute three enclosed open courtyards. With the unsophisticated and narrow entrance, people bow to cross and rise brightly. Trees and buildings interweave silhouetted, elegantly and harmoniously. Minggu Design Agency stationed here fortunately. Objectively, the original conditions of the archaistic "LAI Yard" are not superior, coupled with the lack of time wash-out , its accuracy of process, and the use of material cannot hold a candle to the ancients' handwork. Even so, it still reflects the memory and the desire of people in the present time and space for tradition.

〔来院〕

来，由远到近，由过去到现在，由传统到当代，"来院"由此得名，我们希望在传统的庭院里表达当代。来院的构筑初衷是无组织叠加，可以是一个冥想体验空间，抑或是一个书房，直到项目完成也没有植入任何功能，创作者每天伫立院内，给予原始空间多种状态的想象，一边感知，一边营造。此时的设计变身为一种商谈，一天天内心鏖战，为的是寻找最贴切的答案……

〔交合〕

冥想空间半挑出旧屋基面与庭院交合，原始柱架交合透明围合介质，构筑成外向型封闭空间漂浮于山水之上；内部架构以子母序列构成，颜色对应深浅二系；左右各一间窄室与居中者主次对比，凹凸相映。格局规正，妙趣横生。古与新、内与外、明与暗、传统与现代皆交汇于此，冲撞对比，和谐共生。创作者只表达光和空间，封闭原始建筑除东南方向以外的所有光源，让光线在朝夕之间的自然变化中，通过交叠屋面、序列构架等物理构筑物将虚体光线实体化，而光影随着时间的变化产生不同的角度，空间

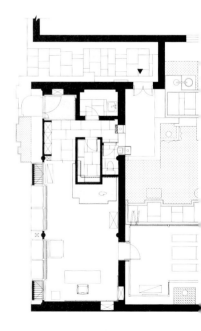

变得让人感动，由"光"将空间呈现，并埋伏"暗"增强空间厚度。仿佛孪生双子，"光"与"暗"彼此勉励、彼此爱慕，又彼此憎恶、彼此伤害。历经暗的挤压，光迸发出更强烈的力量引人深入。院内老井被设置成"地水"之源，通过圆形水器连接折线形水渠将另一端屋檐下收集而来的"天水"汇聚一处，活水流动的路线围合出一池静态山水，将挑出旧屋基面的冥想空间托举而上，院内交通也由此展开。山水纯白，犹如反光板把落入院内的光线温和地送入室内顶棚。创作者在方寸之地步步投射出其二元对立的哲学思考，并企图透过这样的氛围来观察世界的真相。

〔决裂〕

　　设计一定是从功能开始的吗？在商业行为的催生下，越来越多的建筑被赋予功能标签，越来越多的造型行为沦为一种对空间的单纯包装。胡适先生说过，"自"就是原来，"然"就是那样；"自然"其实就是客观世界。创作者固执地坚守着一方不存在商业行为的净地，从美学与环境本身开始建构，坚持院落本身的逻辑关系，不再对所谓瑕疵做浓妆粉饰，功能一直处于一种不确定状态，不再追求均质照明，让光自主营造空间，交还空间表达主张的话语权。与商业决裂的瞬间已无法用语言来拨动心弦。不远处城墙处仿佛蒙着淡淡暗影，带着一丝难以察觉的微笑，气质悲怆仍有渴望。

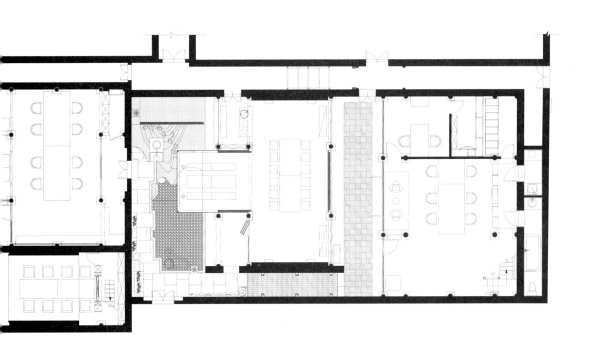

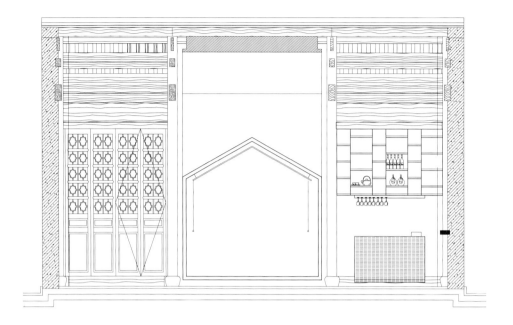

" LAI Yard " is named after " Lai ", which means it is from far to here, from traditional to contemporary. We wish to convey the modern architecture in the traditional courtyard. The original design intention is to superimpose in a disorganized order. It could be a space of meditation or a schoolroom which hasn't been implanted in any function until the architecture is finished. The architecture designer stands in the courtyard every day to show various imagination of the raw space, to architect, and to perceive at the same time. Nowadays, the design has become a chat, and we have inner battles every day in order to seek for the most suitable result.

The meditation area hangs over the base surface of the old house, and interacts with the courtyard. Original column system and transparent material together form an enclosed space, whose facade is facing the courtyard and merging into the Chinese-garden style scenery. The interior structure is inspired by the primary and secondary ranks that correspond to deep and bright colors. The two narrow rooms that are located on left and right of the enclosed space are indented to ensure a sharp contrast to the master room in the middle. The layout of this project is decent, but also full with surprising details. Contrary elements, such as the old and the new, the interior and the exterior, the light and the dark, the conventional and the contemporary, are found to meet here, and merge into a harmony existence. Light and space are the only two themes that the architect emphasized. By blocking all incoming light in the original building, the architect leaves only the southeast side to allow daylight penetration. The vague light becomes tangible and visible through both internal and external roof structure and a series of construction. As time goes by, the light and the shadow also change their angles from morning to evening, resulting in a vivid indoor atmosphere that is presented by "the light" and strengthened by "the dark". Just like a couple who is in a love and hatred relationship, "light" and "dark" also involve nexus between worship and loathing, encourage and damage. In the darkness' perspective, the light appears

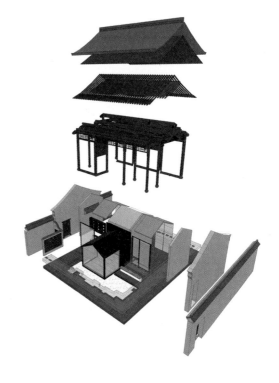

brighter and more attractive. The old well in the courtyard is appreciated as the source of "earth water", which is mixed in a canal that is connected to drains for collecting the "heaven water" from the eaves. The canal leads the mixed water flow along the courtyard, revealing the architecture and the meditation area, as well as deriving the courtyard communication. The bright facade and clear water act as a mirror, sending the daylight from the courtyard to the ceiling inside in a form of gentle reflection. This project illustrates the architect's philosophy and understanding of the dualistic opposition theory, as well as the attempt to observe the truth and reality through such a method.

Does design start from function? Because of the commercial behaviors, buildings are increasingly endowed with functional labels, and more and more modelings descend to a kind of simple packaging for space. Hu Shi once said "Nature" is the exterior world of objective facts. The creator sticks to the non-commercial pure land persistently, starts the construction from the environment, insists on the logical relationship of the courtyard, no longer varnishes the so-called flaws, no longer pursues the homogeneous illumination, allows the light to create space and returns the speech right of space to express proposition. The wall not far away seems to be covered with the slight shadow. It still has the desire with the faint smile and the pathetic temperament.

院内老井被设置成"地水"之源，通过圆形水器连接折线
形水渠将另一端屋檐下收集而来的"天水"汇聚一处，活
水流动的路线围合出一池静态山水，将挑出旧屋基面
的冥想空间托举而上

The old well in the courtyard is appreciated
as the source of "earth water", which is mixed
in a canal that is connected to drains for
collecting the "heaven water" from the eaves.
The canal leads the mixed water flow along
the courtyard, revealing the architecture and
the meditation area, as well as deriving the
courtyard communication.

来院的构筑初衷是无组织叠加，可以是一个
冥想体验空间，抑或是一个书房，直到项目
完成也没有植入任何功能

The original intention of Lai Yard is unorganized
superposition, which can be a space for meditation
or a study. No function has been implanted until
the completion of the project.

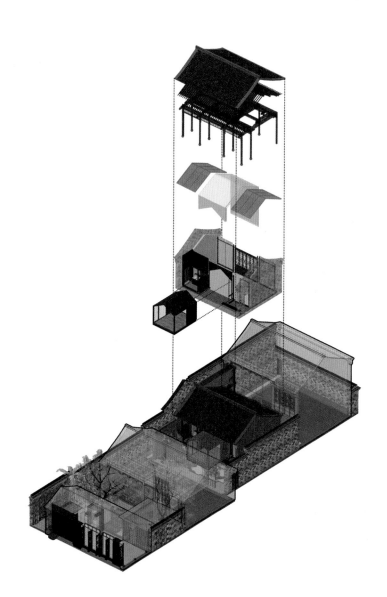

历经暗的挤压，光迸发出更强烈的力量引人深入，内部架构以子母序列构成．
颜色对应深浅二系；左右各一间窄室与居中者主次对比，凹凸相映

After the dark squeeze, the light burst out more intense power to draw
people in depth. The internal structure is composed of daughter and
mother sequence, and the color corresponds to the depth of the two
series. There is one narrow room on both left and right, which is in
contrast with the middle one, concave and convex.

不再追求均质照明，让光
自主营造空间，交还空间
表达主张的话语权

No longer pursues
homogeneous lighting.
Let light create space
independently, and return
the right to express
opinions of space.

"寒夜客来茶当酒，竹炉汤沸火初红。"这是宋代诗人杜耒描写在寒冷的夜里，主人点炉煮茶，以茶当酒待客的诗句。清香茶暖，品茗交谈中其情浓浓，此中儒雅正是宋人传递出的悠悠风韵，是令后世神往的高雅生活。当代浮世尽欢，亦有静心品味当下无边落寞者，竹里馆为此而立。

"A friend came to visit on a cold night. I treated him with tea as the wine. The tea was boiling while the fire was burning red." It is a poem written by Du Lei of the Song Dynasty which describes the scene where the host boiled the tea to treat the guest. The tea brings warmth and aroma, and the two bond with each other in the tea tasting and interaction process. The elegant and refined life is typical of the Song Dynasty which fascinates people. Although the world nowadays is bustling and flashy, there are still people who enjoy the loneliness and appreciate the silence, for which reason the Bamboo's Eatery came into being.

竹里馆

项目地点	南京市江东中路
项目面积	900 平方米
竣工日期	2016.7
主要材料	竹 泥灰 木板
摄　　影	金啸文

Bamboo's Eatery

Project location: Middle Jiangdong Road, Nanjing, China

Project area: 900m²

Completion date: 2016.7

Primary materials: bamboo, plaster, plank

Photograper: Jin Xiaowen

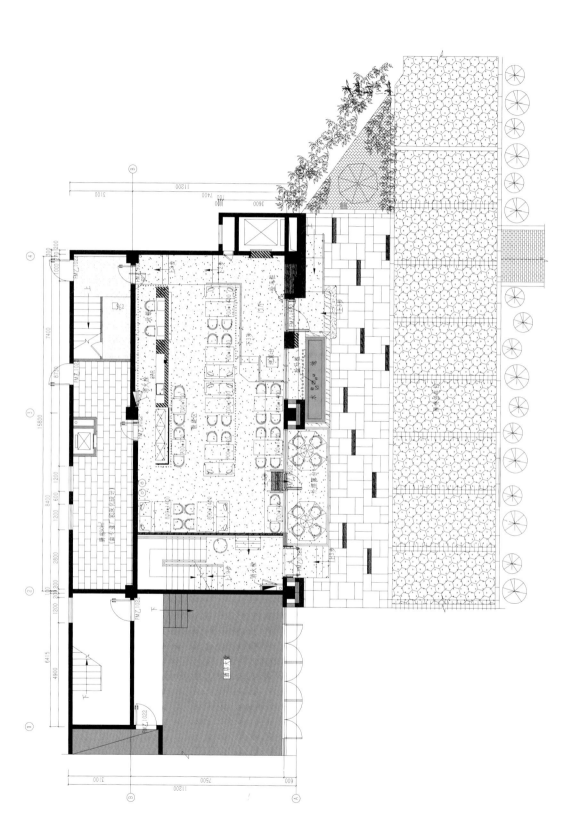

一栋三层临街小楼，以魏晋消散之气为道，喻义君子以白竹为器，尝试一种搭建。搭建似乎更像游离在严肃建筑学之外的民间土木，而搭建带来的空间体验正是将"散"放置在被重新梳理的空间秩序中，这种秩序里最重要的因素——"光"亦是被搭建所带来的"散"重新分解，而获得光线与空间的双重情感，"散"可以告诉你如何塑造弹性的光线。如果说"搭建"是一种放松的尝试，那么梳理则是完整的理性分析。

由外立面的竖向线条延伸至主入口玄关，形成侧向分流进入一层茶歇区。将竹用单一维度的围合方式形成半空间限定区间，茶座布置在竹篱一侧，形成二方连续式的空间关系，并由此聚合成一层的功能核心——"篱园"。此时，围绕着"篱园"的顶面竹篱正发生着维度关系的转变，并引导性地将吧台、出品、服务动线等功能串联起来，与之前的功能核心形成咬合关系，最终指向通向上层的垂直电梯。

通往二层的交通增加了北边的步行体验式楼梯，氧化钢板制作的梯段尝试在有温度的交互中保持部分冷静，从而在进入另一个场域前，以一种旁白的姿态重新整理出独立的情绪。二层茶歇区临窗布置，呈现出较为稳定的状态，入座者更易感受到光线透过窗棂散落桌面的诗话景象。向南的尽头由横竖交织的排竹分割出茶座与电梯厅，向东延伸的排竹将用作洗手功能的饮马槽托举而上，颇有四两拨千斤式的巧力，水源从顶面透过竹管顺流而下，饮马槽的沉重之势被瞬间削减。二层包间区的入口被收纳在一个相对有压迫感的体量内，"压迫"是为了更好地"释放"。在没有自然采光的现场条件下，取西边分割包间与公区的墙面凿壁借光，自然光线在通过茶歇区间后传递到包间内，虽没有斑驳感人的光线落入，却也不失温和透亮，白天被过滤后的光线在相对黑暗的空间内像一张开启光明的网。包间区过道内的墙面除了混合草茎的暖白腻子，亦有七百年历史的城墙砖陈设其中，行走其中可感受时间的穿梭体验。包间内壁留白，取拙朴之姿态，给文人墨客留下足够的臆想与挥毫界面。

三楼全部设置成独立茶舍，依场地东西而立，交通中置，似林中小径，在南北进深三分有二处微微转折。借扭转之态，一个看似溪边草庐的建筑体离地而起，屋檐下探，竹窗由内向外撑起，似乎不论置身内外都有一探窗外究竟的愿望。在狭长的过道中，为获得"静谧中探寻"的行走体验，并有效地将自然光线引入到一个并没有直接对外采光的封闭空间。曾几何时，回忆起很久以前的淳朴年代，门扇没有门套，没有踢脚，却在门扇上方有个邻里孩童打闹时，拴起房门依旧可以翻门而入，被唤作"亮子"的采光神器，可以解决在隔离中交换光线的问题。于是乎，存在于黑暗过道采光面的上部，并由竹篱叠加其中而形成的双层采光界面充当了瞬间解放黑暗的勇士。而顶面转折处被雕塑化处理的局部搭建，正是在空间获得光的解放后所表现出的肆意姿态，有效地软化了相对硬朗的空间对接。包间依旧拙朴、留白。

入座，想起竹林七贤，想起耕读中的陶渊明，也许琴声起时，才是丰满。

将竹用单一维度的围合方式形成半空间限定区间，茶座布置在
竹篱一侧，形成二方连续式的空间关系，并由此聚合成一层的
功能核心 ——"篱园"

The bamboo is enclosed in a single dimension to form a semi-
spatially defined zone, and the teahouse is arranged on one side
of the bamboo hedge to form a two-sided continuous spatial
relationship, from which the "Hedge Garden", the functional
core of the first floor comes together.

向东延伸的排竹将用作洗手功能的饮马槽托举而上，颇有四两拨千斤式的巧力感。水源从顶面透过竹管顺流而下，饮马槽的沉重之势被瞬间削减

The bamboo rows extending to the east lift up the drinking trough for hand-washing, with a sense of harmony and dexterity, as the water flows down through the bamboo pipes from the top, reducing the weight of the trough in an instant.

The Bamboo's Eatery is a three-story complex along the street. It endeavors to revive the spirit of the Wei and Jin Dynasties and adopts the bamboo wood, which implicates the noble man, as the main material to construct the space, which seems to be a civil architecture independent from the orthodox architectonics. The spatial experience brought by the construction process is the embodiment of the "dissipation" philosophy in the rearranged spatial order. The most important element in this order is the light, which is decomposed by the dissipation of the construction and renders a dual sentiment on both the lighting and the space. The "dissipation" tells the secret of how to make the light become elastic. If the "construction" is regarded as a relaxing attempt, then the reorganization is a completely rational analysis.

The vertical lines extend from the outer facade to the hallway of the main entrance, forming a lateral distributary leading to the tea break area on the first floor. The single latitude bamboo wood creates a semi-confined space. The tea seats are arranged alongside the bamboo fence, forming a two-dimensional continuous space relationship and composing the core function area of the first floor — the Bamboo Garden. Now the bamboo ceiling of the Garden goes through a change in latitude, and connects different function areas including the bar area, the production area and the service line, complementing the previous core function area and leading ultimately to the vertical lift to the upper floor.

Another way to get to the second floor is by climbing the stairs to the north. The stairs made of oxidized steel plate show the attempt to retain partially calm in the warm interaction, sorting out independent sentiments again with a voiceover stance before entering another domain. The tea break area is arranged next to the window, presenting a relatively stable state. Guests sitting there will find it easier to appreciate the poetic scene of the sunlight traveling through the window and pouring on the table. The southern end is divided into tea seat area and the elevator area by interwoven bamboos. The arrayed bamboo extending to the east holds the hand washing to sink up, creating an ingenious power against the heavy load. The water runs down along the bamboo pipe from the top, easing the heaviness of the

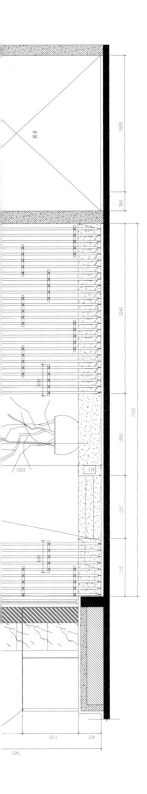

sink immediately. The entrance to the private box area on the second floor is enclosed within a relatively oppressive space, and the "oppressiveness" is for better "emancipation". Without natural lighting, the box area borrows light from the public area by knocking through the wall between the public and the private area. The natural light passes through the tea break area and travels to the boxes, carrying a mild and transparent beauty although not that direct and bright. The filtered light casts a net that brings about brightness in a relatively dark area. The wall surface of the hallway in the box area is painted with white putty blended with grass blade and decorated with wall bricks of 700-year history, making the walk along the hallway a time journey. The interior wall surface is deliberately left unpainted, posing an ancient and primitive touch, leaving free imagination space for guests to taste and create.

The third floor is designed as an independent tea house, standing in the west-east direction. The hallway is designed in the middle of the space, extending like the narrow path in the forest. A slight turn appears at the 2/3 of the north-south passage, where an architectural facade resembling a grass cabin shows up, with the roof overhang leaning to the ground and the bamboo window overstretching outward, attracting the passers-by to peek into the window. In the narrow and long hallway, to obtain the walking experience of "exploring in the silence" and lead the natural light into the enclosed space without direct lighting, the designer employs what is similar to the old lighting apparatus called "Liangzi". Back in old times, there is no door pocket or baseboard on the door leaf, but "Liangzi" was designed to let the light in, and naughty little kids would climb over it to get into the house even when the door was closed. To this end, the double-layer interface decorated with bamboos is designed on the upside of the backside of the dark passage, serving as a bold warrior to liberate the darkness. Moreover, the sculptural treatment of the topo-construction at the turn of the top surface embodies the unbridled stance of the space after being liberated by the light, which softens the relatively stiff spatial transition. And the boxes are still simple and unadorned, leaving sufficient imagination space.

Taking my seat, I think of the Seven Noble Men in the Bamboo Grove, and of Tao Yuanming who read while doing farm work. And when the melodious sound of the Chinese zither comes, the space will be further enriched.

搭建似乎更像游离在严肃建筑学之外的民间土木

The construction seems more like folk civil engineering that strays away from serious architecture.

搭建带来的空间体验正是将"散"放置在被重新梳理的空间秩序中，这种秩序里最重要的因素——"光"、亦是被搭建所带来的"散"重新分解、而获得光线与空间的双重情感

The spatial experience brought by the construction is the "dissipation" philosophy in the rearranged spatial order, in which the most important factor—"light"—is also decomposed by the "scatter" from the construction, so as to achieve the dual emotions of light and space.

光线透过窗棂散落桌面的诗话景象

A poetic scene of light falling through the window frame on the table.

通往二层的交通增加了
北边的步行体验式楼
梯。氧化钢板制作的梯
段尝试在有温度的交互
中保持部分冷静，从而
在进入另一个场域前，
以一种旁白的姿态重新
整理出独立的情绪

Another way to get to
the second floor is by
climbing the stairs to
the north. The stairs
made of oxidized steel
plate show the attempt
to retain partially
calm in the warm
interaction, sorting
out independent
sentiments again with
a voiceover stance
before entering another
domain.

顶面转折处被雕塑化处理的局部搭建，正是在空间获得光的解放后所表现出的肆意姿态，有效地软化了相对硬朗的空间对接

The sculptural treatment of the topo-construction at the turn of the top surface embodies the unbridled stance of the space after being liberated by the light, which softens the relatively stiff spatial transition.

屋檐下探，竹窗由内向外撑起，似乎不论置身内外都有一探究竟的愿望。在狭长的过道中，获得"静谧中探寻"的行走体验

Under eaves, bamboo windows are propped up from inside out. Regardless of whether inside or outside, there is a desire to explore. In narrow aisles, the walk experience of "exploring in silence" is got.

位于万达广场的金元宝食尚汇餐厅以地道的淮扬菜品为主打。对于长江下游扬州至淮安以及周边地区的居民而言，淮扬菜并非大部分国人理解的以刀工和用料复杂烦冗著称高高在上的四大菜系之一。在川菜、湘菜等重口味菜系横行的今天，那是一种家乡的味道。设计师以淮扬大地的现实面貌与百姓生活状态为切入点，携带着淡淡的怀旧情结，似乎童年的回忆一刹那间被唤醒。"印象村野"成了表达餐厅设计的主题，远山、轻风、白云涌动，一片竹篱分割的菜园草地，儿时的竹蜻蜓，貌似是送给了隔壁玩伴？这些活态元素都被抽象化处理成餐厅内的静态造型，等待食客的亲身体验。

印象村野

The Gold Ingot Fashion Food specializes in authentic Huaiyang cuisine. For people living along the lower reach of the Yangtze River from Yangzhou to Huai'an and other nearby regions, the Huaiyang cuisine is not simply regarded as one of the four major Chinese cuisines which is famous for its cutting techniques and the complexity of ingredients as is known to most other Chinese. Rather, it is "the taste of hometown", especially in today's world where the spicy and hot flavor of the Sichuan cuisine and Hunan cuisine are extremely popular. The designer endeavors to reflect the life of people in Huaiyang Region, and instills a slight nostalgic sentiment, arousing people's childhood memories. "Village Impression" serves as the theme of the restaurant. The remote mountains, the breeze, the cloud, the vegetable field with bamboo hedges, and the bamboo dragonfly which I seem to give my huckleberry friend? All these vivid elements are artistically treated to be static sculptures, waiting for the guests to appreciate.

项目地点　　　南京市万达广场
项目面积　　　580 平方米
竣工日期　　　2014.3
主要材料　　　水磨石 砖块 人造藤片
摄　　影　　　金啸文

Village Impression Restaurant

Project location: Wanda Plaza, Nanjing, China

Project area: 580m²

Completion date: 2014.3

Primary materials: terrazzo, brick, artificial vine piece

Photographer: Jin Xiaowen

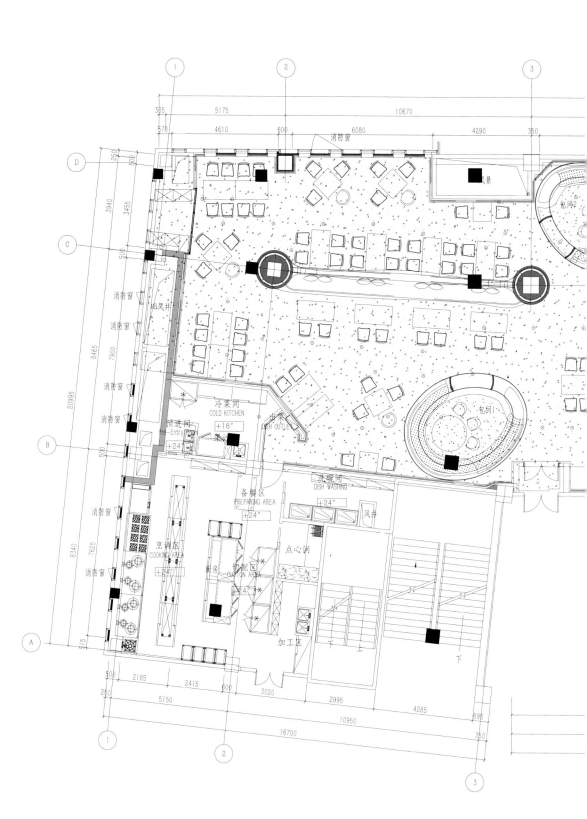

消防窗

消防窗

电风井

消防窗

消防窗

消防窗

冷菜间
COLD KITCHEN
+18°

出菜口
DISH OUTLET

预进间
Pre-Entry
+24°

备餐区
PREPARING AREA
+24°

洗碗间
DISH WASHING

风井

包间1

烹调区
COOKING AREA
+24°

消防窗

消防窗

点心间

厨房
扒配区
CARTON AREA
+24°

加工区

下 上

下

305 5175 10670

570 4610 600 6080 4290 350

250 500

3940

3455

8465

7900

20995

500

8340

7625

515

500 2185 2415 600 3020 2995 4285 695

250 5150 10950 350

16700

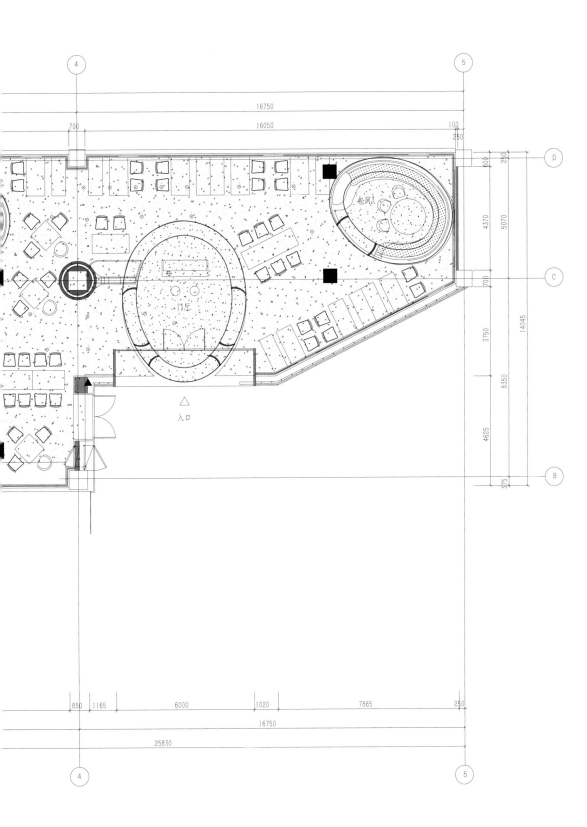

起居室

门厅

入口

16750
16050
700
100
250

250
600
4370
5070

14045

700
3750

8350
4625
375

850 1165 6000 1020 7865 250
16750
25830

4
5
D
C
B

1 1 7

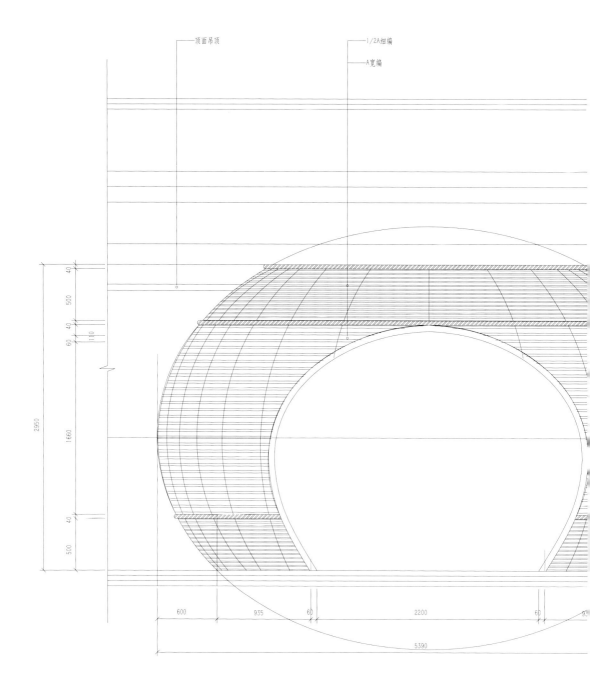

顶面吊顶   1/2A细编

A宽编

40
500
40
60 110
2950
1660
40
500

600   935   60   2200   60   93

5390

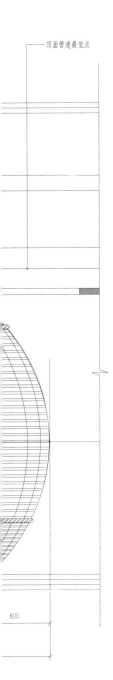

顶面管道最低点

600

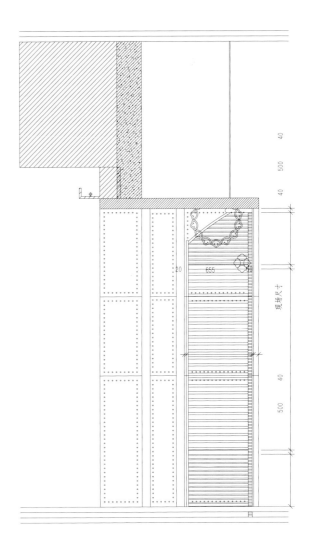

现场尺寸

40
500
40

655
20

40
500

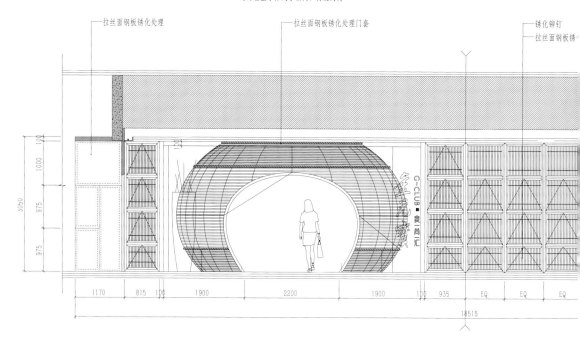

与灯箱上字体大小相同，高度同高

拉丝面钢板锈化处理　　　　　　　　拉丝面钢板锈化处理门套　　　　　　锈化铆钉
　　　　　　　　　　　　　　　　　　　　　　　　　　　　　　　　　　　　　　　拉丝面钢板锈

1170　　815　100　　1900　　　　　2200　　　　　1900　　100　935　　EQ　　EQ　　EQ

18515

　　由蛋形编织体划分出的门厅空间进入，映入眼帘的是门厅背景处被缩影处理的叠加景象，透过一个椭圆的取景框将"远山"呈现出来，食客在第一时间就有一种身处"自然"之中的感觉，正面是接待台，两侧设置进入就餐区的入口，如此，设计师以一种开门见山的方式在蛋形编织体内实现了形象定位、功能设置和交通分流。不足六百平方米的空间内要满足两百人同时就餐，是设计的基础课题。参照现代生活小规模聚餐习惯，就餐单元设计以四至六人为主，长方形餐桌必要时可以灵活搭配。与此同时，三处"巢穴状"半隔断空间可满足淮扬菜系的中国式就餐习惯，以圆桌的方式呈现。乍见平面布局，有种"碧珠落玉盘"的偶发性，那几处"巢穴"也像晒场上的南瓜随意滚落，原本稍显局促的空间被悄悄推启延展开来。同时平面布局和空间组织的不确定性，产生了让人渴望探索的趣味感，新颖的就餐体验也随之而来。看似见缝插针的餐位布局方式，实则是为了节约交通动线带来的空间占有率，"自由"是空间主题，不但体现在平面上，也体现三维空间上。自由曲线的吊顶形式便是苍茫大地上一片涌动的白云，引领着空间内的各种形态走向，灯光的设置，桌椅的布局，形成了一条隐轴，一条无状之状的线索将看似散落的各种部件细致地串连起来，彼此间看似分散随意，却是牵拉着万有引力。

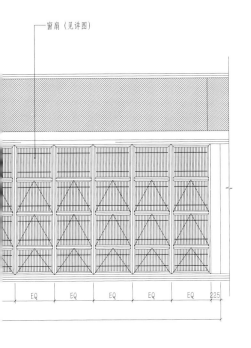

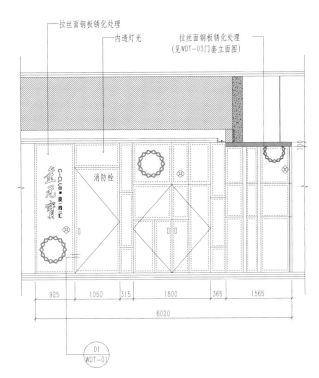

窗扇（见详图）

拉丝面钢板锈化处理

内透灯光

拉丝面钢板锈化处理
（见WDT-03门套立面图）

金元宝

G·CLUB·食尚汇

消防栓

EQ　EQ　EQ　EQ　EQ　225

925　1050　315　1800　365　1565

6020

01
WDT-01

有人说，人生就像一趟列车，我们都坐在中间的车厢，往前走是未来，往后走是过往。一个位子坐久了就总想走动，于是有的人畅想未来，有的人回忆过去。很享受这里的氛围，乡野却不粗野，怀旧却不感伤。设计师没有用所谓"原生态"的粗野材质来植入乡村感的触感体验，而是选择了将现实中的具象转化为表现上的抽象，抽丝剥茧层层蜕变，由抽象思维转变为现实体验，再回到抽象中去。利用温柔精确的表现手法，推敲出一种透着轻松气息的精工细作的潮流感。对于乡野的回忆和向往，被浓缩成一些介乎于抽象和形象之间，为大众广泛理解的形态。功能紧凑，层次丰富，艺术流动。墙面上的远山淡淡地映现，整体浇筑的水磨石地面宁静光洁，映着自家门前的一湾江面。垂直的异型螺旋状灯柱是大堤上旋过的风，随着顶面流云奔走。记忆的现实重组，像叙述一个儿时故事般的叙述空间，把记忆影像转变为体验动线，自己则又成了那个意气风发的少年，放学路上，一路吆喝，呼朋引伴，欢笑玩乐，爬树比赛，不远处的小树林里还有几处秘密"巢穴"，每个少年都有"身兼家国使命"必须要完成的任务。携着回忆的力量，带着感动出发，一些美好，也许回头就看得到，也许前方不远处有更多领悟在等待。

When one walks into the entrance hall in the fabric egg shape, the first thing that comes before his eyes is the epitomized overlay scenery of the "remote mountains" presented via an elliptical "camera aperture". Guests will find themselves walking in the nature. There is a reception desk in the front and on both sides there are entrances to the dining area. The designer realized the functions of image positioning, functional design and traffic divergence within the egg-shape fabric work in a straightforward way. To accommodate more than two hundred people to eat at the same time inside the space less than 600 square meters is the primary concern of this design. Given the small-scale dining habit people have nowadays, the dining unit is designed for four to six people, with the rectangle table designed to serve occasional needs. At the same time, three "nest-like" semi-closed spaces are designed to meet the traditional Chinese eating habits of the Huaiyang region, with round tables arranged within. The layout of the restaurant delivers a randomness similar to that of the pearls dropping on the jade plate. And the "nests" are like the pumpkin dried under the sun, making the relatively constrained space become more extendable in an unnoticeable way. Meanwhile, the uncertainty in the plane layout and the spatial organization spice up the whole space, arousing the curiosity of the guests to explore more and producing novel dining experience. The organization principle of making use of every single space is to reduce the space occupancy of the traffic lines. "Freedom", as the theme of the space, is embodied not only in the plane layout, but also in the spatial arrangement. The free-curved ceiling resembles the drifting cloud over the vast land, guiding different forms and trends. The lighting choice and the layout of the tables and chairs form an invisible axis which joins together the seemingly scattered parts in a clue without a certain form. The different parts seem to be disconnected, but are actually driven by a common force.

Some people believe that life is like a train journey and we all sit in the middle coach. If we go forward, we walk into the future and backward, the past. People tend to move around if having stayed in the same place for too long. Therefore, some people are expecting the future while some are lingering in the past. I enjoy the atmosphere here, which delivers the village impression which is not vulgar and rustic, and which is nostalgic while not sorrowful. The designer did not employ the so called "primitive and cruel" materials to convey the

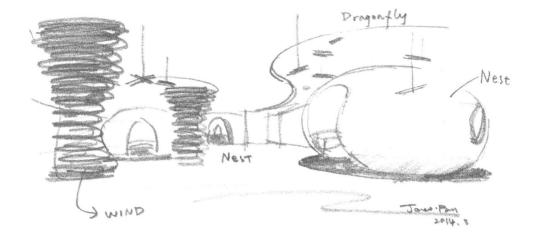

几处"巢穴"像晒场上的南瓜随意滚落，原本稍显局促的空间被悄悄推启
延展开来，同时平面布局和空间组织的不确定性，产生了让人渴望探索
的趣味感

Several "nests" are like pumpkins randomly rolling down from a drying field, leaving the formerly slightly confined space open and extended; meanwhile, the uncertainty of the layout and organization of the space creates an interesting sense of desire to explore.

countryside experience, instead, he transforms concrete things in real life into abstract expressions. And with painstaking efforts, the abstract expressions are again transformed into real-life experiences, then back into the abstract forms. The designer employs gentle and precise expression techniques and produces a sense of fashion which is relaxed and refined. The memories and longings for country life are epitomized into some specific forms in between the concrete and the abstract, which are commonly recognized by the general public. The space has a compact function design, with rich layers and artistic mobility. The remote mountains on the wall look light and pretty, and the one-piece casting terrazzo is neat and serene, reflecting the river running before the front door. The spiral vertical lamp stand resembles the wind wuthering over the dam, traveling along with the drifting cloud. The concrete realities of the childhood are reconstructed here, unfolding a narrative space like telling a childhood story, turning memories and images into dynamic experience. People here become the little proud boy again, calling out all friends on the way back home from the school. Laughing and playing, they join a tree-climbing competition and there are several hideouts of them in the nearby forest. Every one of them has the dream to be a superhero to save the world. Empowered and touched by childhood beautiful memories, one is prepared to forge ahead. The beautifulness is still right there, waiting to be picked up while unknown journey awaits on the road ahead.

Next

white clouds

Windy

设计师以一种开门见山的方式在蛋形编织体内实现了形象定位、功能设置和交通分
流 由蛋形编织体划分出的门厅空间进入，映入眼帘的是门厅背景处被缩影处理的叠
加景象，透过一个椭圆的取景框将"远山"呈现出来

The designer realizes image positioning, function setting and
traffic diversion in the egg-shaped weaving body in a straight-
to-the-point way. Entering the foyer space divided by the egg-
shaped woven body, what comes into view is the superimposed
scene of the miniature processing in the foyer background.
The"distant mountain"is presented through an oval frame.

墙面上的远山淡淡地映现，整体浇筑的水磨石地面宁
静光洁，映着自家门前的一湾江面，垂直的异型螺旋
状灯柱那是大堤上旋过的风，随着顶面流云奔走

The distant mountains on the wall are reflected
faintly. The whole terrazzo floor is quiet and clean,
reflecting a bay of the river in front of your house.
The vertical special-shaped spiral lamp post is the
wind swirling on the levee, running with the clouds
on the top.

"巢穴状"半隔断空间以满足淮扬菜系的中国式就餐习惯,
以圆桌的方式呈现

"Nest-like" semi-partition space is designed to
meet the Chinese dining habits of Huaiyang
cuisine, presented in a round table.

一下雪，南京就变成了"建康"。戊戌年末，大雪如约而至。此时离居住样板面市已两月有余。不出所料，市场给予了积极正面的反馈，溢美之词不绝。再见此宅，惊异于其在隆冬中所展现的另一种气质。冬与春所得到的体验是那么的不同，在这里空间与时间维度都保持着延展的弹性。

即便使用的是现代构造技艺工法，仍能深刻体味到江南气息。是的，对于周边文脉的协调问题，我们从来不缺乏答案。模仿、装饰性的怀旧带来的仅仅是一种表面上敷衍的结合，古典平庸的抽象化、虚构一个老旧的意识形态完全没有办法将哲人气质进行传递。基于对地方历史的理解，对重量、结构和装配的理解，创作者移情于建筑，使用标识化的节点做法，将抽象形式的元素从回忆中剥离开来，以精简成生活的本质。

长
物
之
宅

Nanjing became "Jiankang" as soon as it snows. At the end of the Wuxu, heavy snow arrived as scheduled. At this time, it has been more than two months since the housing model came on to the market. As expected, the market gave positive feedback, full of praise. When seeing this house again, I was amazed at another different temperament it showed in the dead of winter. The experience of winter and spring you can get from here is so different, where the dimensions of space and time maintain extended elasticity.

Even with modern construction techniques, you can still deeply get a taste of Jiangnan atmosphere. Yes, for the coordination of circumjacent context, we are never short of answers. Imitation and decorative nostalgia brings only a perfunctory combination apparently. The classical mediocre abstraction and the fiction of an old ideology has no way to convey the philosophical temperament. Based on the understanding of local history, of weight and structure, creators emphasize on architecture. They use identified node approach to separate the abstract forms from memories, to compress into life essence.

建 筑 设 计 团 队	GOA 大象
景 观 设 计 团 队	北京顺景园林
室 内 及 庭 院 设 计	名谷设计机构 & 东琪及道
特 约 艺 术 家	杨勇
项 目 面 积	A 户型 320 平方米
	B 户型 350 平方米
主 要 材 料	氟碳钢板 葡萄牙米黄大理石
	木饰面 烤漆板
竣 工 日 期	2018.12
摄 影	李国民 金啸文

Residence of Zhang Wu

Construction design team: GOA Elephant

Landscape design team: Beijing Shunjing Garden

Display design & execution: Minggu Design Agency,
Dongqijidao

Special guest artist: YangYong

Project area : Unit Model A, 320m²
Unit Model B, 350m²

Main materials: Fluorocarbon steel plate, portuguese
beige marble, wood finish, lacquer board

Completion date: 2018.12.

Photographer: Li Guomin, Jin Xiaowen

冬与春所得到的体验是那么的不同，在这里空间与时间
维度都保持着延展的弹性

The experience of winter and spring you can get here
is so different, where the dimensions of space and time
maintain extended elasticity.

〔九月森林〕

　　宅有三院，以南院最为明媚。水面、草地、石板以几何形状写意组合。水中有桥、桥有廊、廊有亭，一派萧散简远。尤其以亭的构建为妙，屋面将结构、遮雨、柔光三层关系分离处理，两排格栅、两组圆柱纵向呈对角布置，释放阳角压力，赢得一片轻松。石墙沉默，涟漪闪烁，独坐幽篁，是非无处落脚，忧喜自上眉梢。水面如镜将风卷云舒、日月星辰、远山魅影一一纳入这方圆。亭中挂有大型植物画作一幅，与周边写意植物形成了强烈的对比。这并非随意触发，我们有理由怀疑这甚至包含着创作者有意造作的成分。生命之境界有隔，我需要一些略显可爱的世俗满足感。于是，便呈现出如此的率意而为。

　　北院入口谦逊，道路转折，雨篷微微下压，高大植物衬于雨篷后、立于天井中。高、中、矮植物分级种植，漫步石径，树叶轻抚身侧，才入家门旋即感受到这份温柔湿润。此时，晚归者映于白墙上的剪影被忙碌的厨娘尽收眼下，透过卷轴般横向展开的窗棂，一时竟难以分辨谁才是画中之人。专业化动线的厨房在家庭使用中很少遇见。出入分置，入口结合门厅，出口面朝餐厅。创作者用一组"空间面纱"序列，利用视觉完美法则，在客厅餐厅部位取得了相对完整、稳定的完美空间。此空间的透明边界又与南院呈现出一种无缝状态，模糊了时间与空间的界限，内外空间联系的增强，将人们从建筑有限的内部空间中释放出来。

　　在艺术画作的引领下，拾级而上，到达二层居住区。主卧伺服空间中置，电动雾化玻璃将私密性和通透性切换自如。"空间面纱"再一次粉墨登场，不同功能空间之间的功能透析以及非透明状态下的交通问题都由其承担解决。以歙砚制作而成的洗手台盆、隐藏式的下水线路处理、冲淋房阴角弧形处理，以及拼接的考究等，处处都呈现出对生活细节的推敲。三层面积不大，趣味点不少。首先由

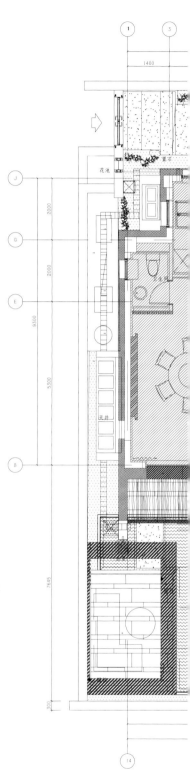

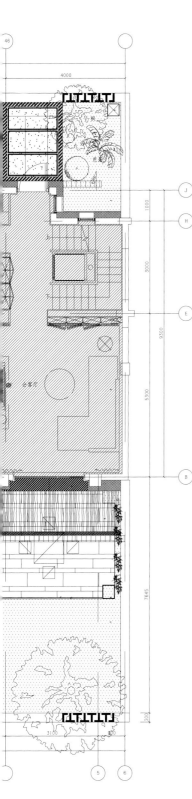

一种度假式居住理念提出一种解决小空间使用的方案，房间与房间之间可分合，户外与室内可分合。电梯厅顶部与原有屋面层剥离开来，形成新的天井，四周缝隙将自然光引入并挤压成束，落于光线昏暗的墙体饰面上，神秘感油然而生。使用者在与顶部光线的互动中，产生了对天光方向的强烈向往。

除去南北，还剩一院，翻遍平面也难寻到，人多称之为"天井"，创作者宁可多呈一份尊重，名为"光院"。尽管尺度较小，光院很好地完成了对地下天光的输送及空间气息的循环工作。通过将伺服空间中置，地下空间自然分裂成交通、休闲两大区域。交通与车库贯通，休闲与光院相接。最简练的排列组合营造出带有清爽感的平面功能构成。

〔结庐三姿〕

食人间烟火，而后有居。抒发情感，而后生活。土地厚热，下探的屋檐，从庇护开始，到仁爱结束，居住方式的探讨从未绕开时代。

现代主义建立的模块式特征，最初是解决劳工阶层的生存居所，此后，由演变而获得再生的公寓住宅，承担起亿万民众的生存寄托。除去解决情感问题，其实用性几乎无所不能，并由此承载出设计的机遇与价值，由需求引发的市场化建造表现为直接、速度、局部、复制、趋势感，总结为被消费。"消费"是"建造"服务"居住"的驱动源，人类最基本的居住层级，"消费式居住"在城市化进程的蔓延中无处不在。

居住绕不开时代，更绕不开时间，风土习俗无形中掌控着一切。当建筑者披着传统或当代的外衣，手握现代或后现代的利剑，建造心中的理想国，建筑者的欲望可以瞬间描绘一个宏大的叙事。而生活只在意爱与被爱的点滴，不卑不亢地与时间相守前行，生活中被检验的人与人、人与人造物之间的矛盾，被阶段性激发显现，并非像建立"消费式居住"那样，在建造之前，以站在时间轴中间段的立场，去总结时间轴，忽略生活随时间变化的生长感，回避现实与理想的差距。生活中，允许新人物的加入，对物理空间产生的对抗与包容，允许人物性格的改变，对场域做出的调整余地，允许季节变化对日常习惯的干扰，"生活式居住"更尊重人性的通达和建造的想象力。品味当下，将欲望变作一粒美好的种子，播向生活的冗长。

中国道家哲学的"无为"与西方现代主义建筑观念"少即多"在某个维度上是相通的，如果说每个人的生命长度是一个恒定的时间，那么我们每过一天，都是在人生的长度上做出减法，不是吗？生活的欲望让我们饱含热情，对未来憧憬，积极回应生存矛盾，但欲望的反面呢？一个永远无法满足的物欲横流，得到越多负担随之，新的"人－物"矛盾需要解决，就像一个大房子，需配置更多的人力去维护清洁，院子里的花草需精心养护。我们建立一个个看似美好安全的空间壁垒，心灵却难以安放。其实大自然早已给予了一切，清晨的露珠，在朝阳的映射下晶莹剔透，阳光穿过庭院的嫩绿，透过窗棂的影，洒落在书案的纸页间，榻前清茶读书格物，纵情山水斗转星移。热爱！来自天地的变化，岁月的痕迹，心性的磨砺。"情感式居住"像是阅尽世间繁华的归者，冷眼旁观生活信众相互追逐的影子，对生活最本真的部分心存敬畏。

〔长空琐记〕

对于每一个人来说，总有那么一位导师，他的出现似乎比其他导师都更为重要。于我而言，长物之宅的创作者勉强算是这样一个人吧。或许说不上极具魅力，但却是充满了吸引力。他能够哀时抚琴、醉酒当歌，当然包括探索觉知的入口以及揣摩具象的形态……

如文震亨所绘，晚明时期文人恋物超然物外，寄情于物，终究恋上的却非物趣本质，渴求凸显的恰是文人其自身之高雅气质，从而达到物我合一的状态。

是居住建筑就始终摆脱不了其住所本质带来的功能目标，自然也不能单纯以空灵格物的审美来品调。然"作为高级别住所，在满足了生活、消费两大基本功能后，是需要我们把大规模的注意力集中

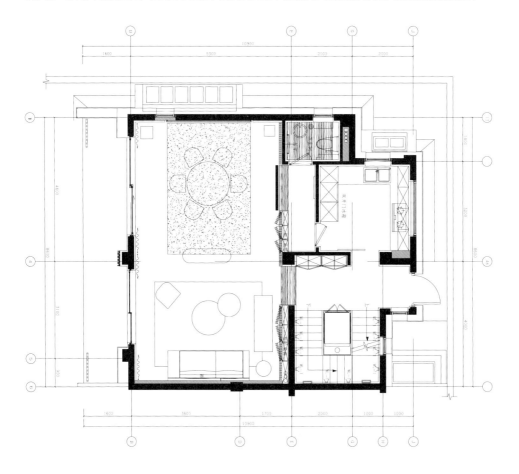

在其情感需求上的"，创作者数次强调，"建筑者并非空间的主导者，我们需要与使用者一同在建筑中游走"。这种追求满足身体经验的意图非常明显——追求不能语言文字化的东西，借"物"来脱离语言上的建筑主义，用体验来描述空间。以物明理，用形而下来传达形而上本身就是一种极具挑战性的行为。只因独特体验来源于不同个体，在造空格物过程中自身感受对象给予的反哺。

September Forest

There are three houses in a residence, and the south courtyard is the most bright and beautiful one. Water surface, grassland and flagstone are combined in freehand geometrical shapes. The bridges in the water, galleries on the bridge and pavilions on the gallery show a style of "superior to mundane affairs". In particular, the construction of the pavilion is the best. The roof separates the relationship among structure, rain shelter and soft light. Two rows of grids and two groups of cylinders are arranged diagonally in the longitudinal direction, to release the pressure of out corner and to win a piece of ease. The stone walls are silent and the ripples are twinkling. Sitting alone, without the trouble of right or wrong, you can sense the sorrow and joy come from the top of your brows. The surface of the water is like a mirror which brings the wind, the sun, the moon and the stars, as well as the phantom of the distant mountains into this circumference one by one. There is a large plant painting hanging in the pavilion, which forms a strong contrast with the surrounding freehand plants. The contrast is not triggered randomly. We have grounds to doubt that it even contains the compositions that deliberately created by the creator. "There is always a gap in the realm of life, and we need some mundane satisfaction that seems slightly lovely." As a result, it presents the work like this, frankly and naturally.

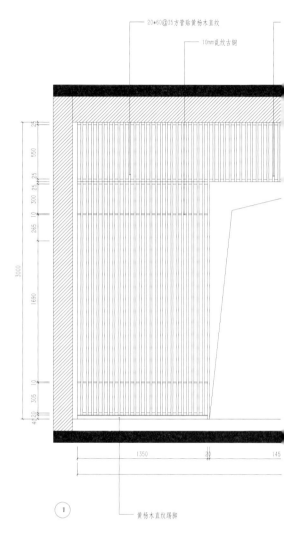

The entrance to the north courtyard is humble, the road turns, the awning presses slightly, and the tall plants stand behind the awning and stand in the patio. Plants are planted according to "high, medium and low" height. Walking on the cobblestone path, the leaves gently touch you to say hello, and you can sense this gentle and humid feeling as long as you walk through the door. At this time, the silhouette of the late comer reflected on the white wall is taken in by the busy cook. Through the window lattice spreading horizontally like a scroll, it is difficult to tell who is in the painting for a while. Professional kitchens are rarely encountered in family use. Exit and entrance are designed separately. The entrance is combined with the livingroom, and the exit faces the diningroom. The creator uses a set of "space veil" sequence and uses the law of visual perfection to obtain a relatively complete and

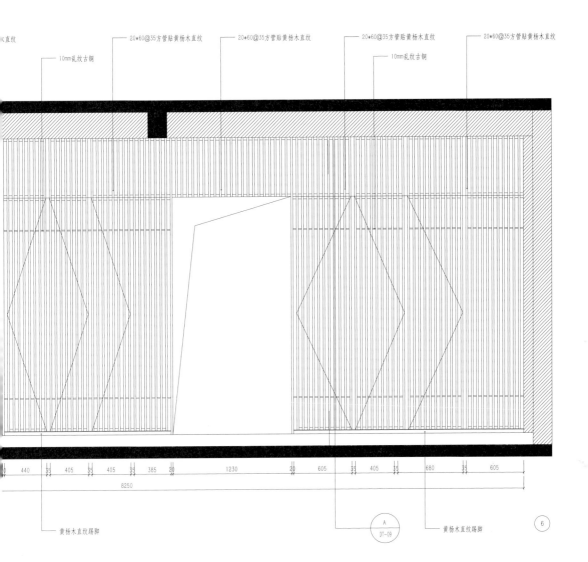

0.直纹
10mm乱纹古铜
20•60@35方管贴黄杨木直纹
20•60@35方管贴黄杨木直纹
20•60@35方管贴黄杨木直纹
20•60@35方管贴黄杨木直纹
10mm乱纹古铜

| 440 | 405 | 405 | 385 | 1230 | 605 | 405 | 680 | 605 |

8250

黄杨木直纹踢脚

A
DT~09

黄杨木直纹踢脚

6

stable perfect space in the living room and dining room. The transparent boundary of this space presents a seamless state with the south courtyard, blurring the boundary between time and space. The enhancement of the relationship between internal and external space releases people from the limited interior space of the building.

Under the guidance of art paintings, you can follow the steps to reach the second floor to the residential area. The servo space of the master bedroom system is configured in the middle, and the electrically atomized glass can switch the space freely between private and public. Once again, the space appears on the stage, and it is responsible for solving the functional dialysis between different functional spaces and the traffic problems in the opaque state. The sink basin made by She Inkstone, the hidden water line processing , the arc treatment of the internal corner in the shower room, and the exquisite splicing all show a careful consideration of the details of life. The area of the third floor is small but full of fun. First

of all, it puts forward a solution to the use of a small space with a concept of holiday-style living. Each room can be intercommunicated or isolated, and the outdoor and indoor rooms can also be intercommunicated. The top of the elevator hall is separated from the original roof layer to form a new sky well. The natural light is introduced and squeezed into a beam by the gap around the elevator hall, which falls on the wall surface with a dim light, and the mysterious feeling arises spontaneously. In the interaction with the top light, users have a strong yearning for the direction of the sky light.

Apart from the north and south yard, there is still one courtyard left, and it is difficult to find it all over the plane. Many people call it "Patio", and the creator would rather show more respect, called "Light Courtyard". Although the scale is small, the "Light Courtyard" has well completed the transportation of underground skylight and the circulation of space breath. By setting the servo function in the middle of the room, the underground space is naturally divided into two major areas: traffic and leisure. The traffic area is connected with the garage, and the leisure area is connected with the light courtyard. The most concise arrangement and combination creates a plane functional composition with a sense of freshness.

Three kinds of artistic conception of residence construction
After satisfying their appetites, human beings urgently seek a safe place to live. Then their emotional needs are met. That's life. From the deep soil to the overhanging eaves, as from the shelter to the love, the discussion of the way to live has never been independent of the times. Modernist architecture, with its modular features, was originally designed to address the subsistence of the working class. Since then, the regenerated apartment, which is obtained from the evolution, takes the responsibility of the survival sustenance of hundreds for millions of people. Except for solving emotional problems, the practicability is almost omnipotent. Thus, it carries out the opportunity and value of design. The market-oriented

construction caused by demand is presented as direct, speed, local, replication, and a sense of trend. In a word, it is summed up as being consumed. "Consumption" is the driving force of "construction" and "residence". As the most basic level of human habitation, "consumption housing" is everywhere in the spread of urbanization.

Living can not go around the times, not to mention let alone time. Local customs virtually control everything. When the designers wear the cloak of traditional or contemporary, with modern or post-modern swords in their hand, building the utopia in their hearts, their desire can instantly depict a grand narrative chapter. However, life only care about love and being loved, being counted with time going on, neither haughty nor humble. The contradictions between people and people and between people and artifacts which is being tested in life are aroused and revealed by phases. It is not like the establishment of "consumption housing". Before construction, they summarize the timeline from the standpoint of standing in the middle of the timeline, ignoring the growing sense of life changing over time, and avoiding the gap between reality and ideal. In life, it is allowed for new characters to join, including the confrontation and tolerance of physical space. The characters are allowed to change, to make a room for adjustment for the field. The interference of daily habits which are led by seasonal variation is also allowed. "Life-style living" respect more of the understanding of human nature and the imagination of constructions. Savor the present, turn desire into a beautiful seed and sow it into an existential life.

The Chinese philosophy of "inaction" is similar to the western modernist architectural concept of "less is more" in a certain dimension. If the length of each person's life is a constant time, then each day we spend is a subtractions on the length of life, isn't it? The desire for life makes us full of enthusiasm. We are looking forward to the future, and actively respond to the contradictions of existence. But on the opposite side of positive desire is negative desire, which is a crazy material need that can never be satisfied. The more we got, the more burden follows. Meanwhile, the new "human-material" contradiction needs to be solved. Just like a big house, it needs to allocate more manpower to maintain cleanliness. The flowers and plants in the yard also need to be carefully maintained. We set up space barriers that seem beautiful and safe one by one, but it is difficult to find a space to place the soul. In fact, nature has already given everything, the dew in the morning is glittering and translucent under the reflection of the rising sun. Sunshine passes through the tender green of the courtyard and the shadow of the window lattice, sprinkling on the pages of the book. Read books in the morning with a cup of tea, appreciating how time flies on the landscape. This is love from the changes of heaven and earth, the wisdom of the ages,and the grind of mindset. "Emotional living" is like a person who has seen all the prosperities in the world, watching coldly from the sidelines the shadows of human beings chasing around each other, and still hold the discipline in awe of the most authentic part of life.

The miscellaneous notes of Changkong

For everyone, there is always a mentor whose presence seems to be more important than any other mentors. As far as I am concerned, the creator of The Residence of ZhangWu can barely be regarded as such a person. He may not be that charming, but full of attraction. He

can plan the Gu Zheng in times of mourning, get drunk and sing, and explore the entrance to awareness and figure out the form of the concrete morphology.

As Wen Zhenheng painted, the literati in the late Ming Dynasty loved things beyond transcendence and fell in love with things, but what they eventually fell in love with was not the essence of material interest, but their own elegant temperament and the state of unity of object and oneself.

Residential buildings can not get rid of the functional goals brought by the essence of their dwellings, and naturally they can not be characterized simply by the aesthetic appreciation of ethereal objects. However, as a high-level residence, after meeting the two basic functions of life and consumption, it needs to be focused on people's emotional needs. The creator stressed several times: "the designer is not the leader of the space. We need to walk in the building with the users." The intention of this pursuit to satisfy physical experience is very obvious—to pursue things that cannot be written in language, to break away from linguistic constructivism by "things" and to describe space with experience. It is a challenging behavior to convey metaphysics in terms of things and metaphysics.

电梯厅顶部与原有屋面层剥离开来、形成新的天井，四周缝隙将自然光引入并挤压成束，落于光线昏暗的墙体饰面上，神秘感油然而生

The top of the lift lobby is separated from the existing roof to form a new atrium, with gaps around it bringing in natural light and squeezing it into beams on the dimly lit wall finishes, creating a sense of mystery.

亭的屋面将结构、遮雨、柔光三层关系分离处理，与纵向两排格栅、两组
圆柱呈对角布置，释放阳角压力、赢得一片轻松

The roof separates the relationship among structure, rain shelter and soft light. Two rows of grids and two groups of cylinders are arranged diagonally in the longitudinal direction, to release the pressure of out corner and to win a piece of ease.

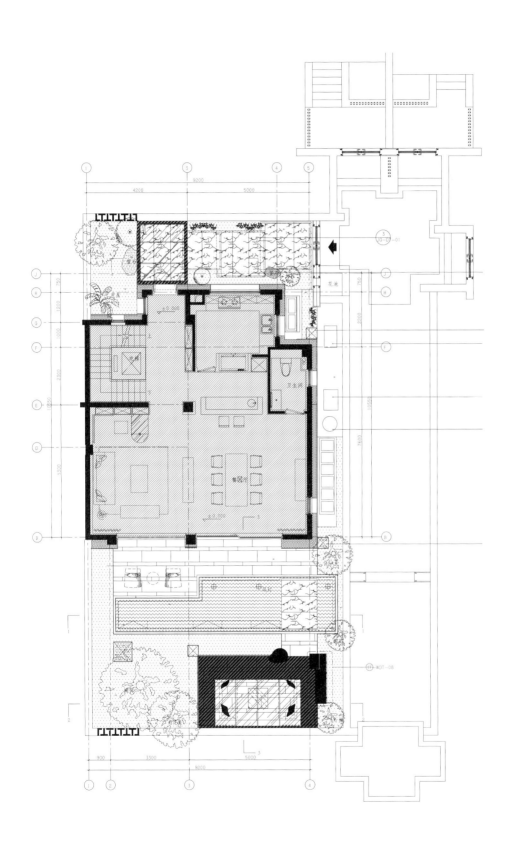

内外空间联系的增强，将人们从建筑有限的内部空间中释放出来

The enhanced spatial connection between the interior and exterior releases people
from the building's limited interior space.

地下空间自然分裂成交通、休闲两大区域。交通与车库贯通，休闲与光院相接

The underground space is naturally divided into two major areas: traffic and leisure. The traffic area is connected with the garage, and the leisure area is connected with the light courtyard.

如果"营造预想"是对设计最初浅的解释,对于民众人居是功能性帮助,那么空间的觉知,可以被认为是众多生命体验中最为磅礴的一课。古人有气吞山河者、有胸怀天下者、有移竹当窗分篱为院者,今人有摩天大楼与微信朋友圈,皆磅礴,皆空间。灵魂与身体安放处,觉察感知种种,尚古或维新,繁华过后总有未明之体验,此种未明体验正是空间觉知的入口。当代设计者,以建筑空间为媒介,揣摩未明体验所营造出的具象形态,正是吾辈前赴后继为之努力的应有姿态。

自先秦之后,中华智慧逐渐参与并深入的自然觉醒中,"二十四"冥冥之中成为一个特殊的数字,熟悉且神秘。季节二十四节气、太极二十四式、风水二十四山法…… 无穷变换,不离本源。这正是餐厅创始人思考美食制作的灵感来源,用现代料理的方式将古老的淮扬食味重新演绎。"二十四单"的生长过程似乎并非单纯的建造,旧厂房改造空间窄小,自然采光杂乱等都只是寻常外因,而赋予空间血肉性格,建立新的未明体验,从多维的角度启发参与其中的人的觉知感受,是本次营造所面临的新课题。

Designing, if understood preliminarily as the process "to construct what one expects", mainly serves for functional needs. However, as a way to perceive the space, it is one of the most enlightening lessons for life. In ancient times, there were people who were ambitious and daring to conquer the mountains and rivers, who were patriotic and caring to seek welfare for the whole world, and who were refined and reclusive to live with the bamboos planted as the window and the pear trees standing separately as the yard. Nowadays, we have skyscraping buildings and WeChat "Moments". All these are related to one concept—the space. It is the space that provides shelter for our body and soul. We discover and perceive things, and learn to respect the past or value the present. But always we find uncertain experience which leads us to the entrance of space awareness. In modern times, the designer should endeavor to use the architectural space as the media to explore the concrete forms generated by the uncertain experience.

Ever since the pre-Qin period, the Chinese wisdom is embodied in the awareness toward nature. The number "24" became a special figure which carries familiarity as well as mystery. We see various expressions of 24, including the 24 solar terms, the 24 Tai Chi moves, the 24 laws of Fengshui and so on. All of them have a common core, which is also the origin that drives the owner of the restaurant to cook dishes to revive the time—honored Huaiyang Cuisine in modern cooking techniques. The process to design and build the restaurant is not a simple construction process and the designer faces many challenges, including the normal expected ones such as the reconstruction of the old factory, the narrowness of the space, the disorderliness of natural lighting and new specific ones such as how to give character to the space, how to construct new uncertain experiences and how to arouse the awareness of the guests in a multidimensional way.

全息的觉知 · 二十四单

项目地点　　南京市来凤街
项目面积　　1 700 平方米
竣工日期　　2018.3
主要材料　　木 涂料 石材
摄　　影　　李国民

24 Cathay Restaurant

Project location: Laifeng Street, Nanjing, China

Project area: 1 700m²

Primary materials: wood, paint, stone

Project completion date : 2018.3

Photographer: Li Guomin

〔工事〕

　　一两个世纪前，这里曾作为铸币厂，在国民时期显赫一时。在沿东西向轴线坐落层高十余米的红砖建筑单层厂房内，脱离原建筑外墙，建立新的空间秩序，规划出两层半内部空间，受原建筑外皮庇护同时，结构上保持相对独立。主入口东临来凤街，回应相对市井面貌的外部环境，有必要不动声色地做出气质转换。为此，由纤细圆管制作的抽象"竹林"从户外绿篱向室内延伸，透明材质分割出内外界限并放置在"林"中，被模糊的界面使新生空间的生长变得顺理成章。

　　由林入厅，呈现出双层挑高尺度的错落关系，这种关系是体察原有建筑内部空间后的理想塑造，再将"功能性"放置在理想塑造后的美学形体内，产生出层级交通动线，如此，由"觉知"引发"美学"塑造。再植入"功能"最后形成"交通"的逆向营造逻辑，更具牵引力。高尺度厅堂存在的意义，是为自然空间向人造空间的过渡做好准备。

　　通往半地下一层的空间被划分成开敞式厨房与就餐区，食客与制作者的关系应是互为赏识的，前提是保有距离，一种互通并不断保持变化的远近关系。黑色条形钢板制作的区间分割面被设计者称为"空间面纱"的序列，在食客行走的过程中，视线不断被放大，发现食物制作的过程当是欣喜。"觉知"此刻表现为互赏者的心理交流。就餐区的临窗立面，开窗杂乱之际光的肆意，用宣纸材质裱糊出的屏风墙，将自然光线进行二次过滤，晕染出糯白的儒雅气氛，黑与白的冲突美学，一如面纱的正反面。

　　沿厅右侧拾级而上，到达二层私密就餐区，通道中置，两侧各为大小私密就餐单元。为促进食客快速通过，避免公区停留，黑色木饰面材质制作的轻薄界面由顶面包裹至立面，并将各单元相互串联，呈现出相对压迫且神秘的气氛，白色石材地面配合灯光序列，成为引导进入的显性指示。

　　由通道立面设置的隐框门进入，忽然开朗之余更多的是面对隐逸环境的兴奋与自信，临窗界面依旧采用宣纸屏风将自然光线过滤并充斥空间，光线照射后的树影在某个时段内成为主角，与佳肴一同演绎。

〔惊蛰〕

　　师法从宋元，神思入晋唐，今人当在古意中寻造化。西方人的生存方式很直接，很现实，不绕弯子，建筑空间要好用，首推功能性，现代主义兴起后更是将"功能"的意义教科书化，以致于在相当长的一个时期内，在中国本土设计语境中，也是全民歌唱"功能"的美妙。其实"功能"远没有那么强大，强大到可以通过类似冲突运算的方式，去解决人与人造物之间的矛盾，一个舒适并拥有即时体验的空间，是多维度信息传递的结果，所谓解读空间，正是探索空间的觉知力使然。这种觉知充满了冒险家精神，一面攀向高尚的顶峰瞭望更为壮美的风景，也会有不慎跌落低谷后的慌乱，疲于应付失去逻辑的冲突运算，前者赋予设计营造的结果是突破与惊喜，后者只能是被动地接受，在应接不暇的压力催生下将营造变得低劣不堪。

这里曾作为铸币厂，在民国时期显赫一时

The place was famous as a mintage spot during of the Republican of China.

The Construction

More than a century ago, this place was famous as a mintage spot in the republican period. The red –brick construction lies in the west-east direction and was originally a one floor factory of about 10 meters high. The designer broke the original form of the space and constructed a new spatial order, turning it into a space of two and a half floor architecture conforming to the former outer appearance while retaining its independent structure. The main entrance faces the Laifeng Street to its east, which agrees with the overall civil atmosphere while unnoticeably undergoes temperament adaptation. To this end, the concept bamboo made of thin circular tube extends from the outdoor fence space into the exterior space, with transparent materials designed as the dividing line of the indoor and outdoor space. The blurry interface makes the growth of the newly built space natural.

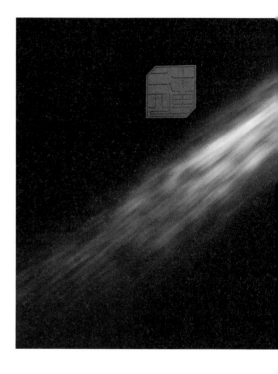

Walking from the bamboo grove to the dining hall, the guest is introduced to the two-floor high ceiling space of uneven relationship. This relationship involves the ideal shaping which respects the original interior space and the implantation of " the functional needs " inside this aesthetic form after the ideal shaping, producing a multi-layer dynamic traffic line. In this way, the "space awareness" leads to " aesthetic creation ", which, combined with the functional concerns, produce the final " traffic " of the space. This reverse thinking logic has more tractive force. The high ceiling hall prepares the transition from the natural space to the artificial space.

The space leading to the semi-underground is divided into two functional areas, the open air kitchen area and the dining area. Guests and chefs are supposed to share a mutual appreciation, with the premise of keeping a certain distance from each other. The distance embodies a relationship which preserves the connectivity while keeps changing all the time. The dividing interface made of black steel bars and the sequences which is called by the designer as "the veil of the space " all come closer to guests as they walk further with their vision getting broader. Finally guests will feel great pleasure as they come to see the production process of the food. The " awareness " here comes as the spiritual communication between the appreciators. The window-side facade receives randomly-casted sun light when the window opens. The rice-paper texture screen is able to filter the natural light for a second time, creating refined and modest atmosphere. The conflicting aesthetics of the white color versus the darkness, just as the inside and outside of the veil, creates unique beauty.

Climbing the stairs on the right of the hall, one enters the private dining area on the second floor. There is a passage designed in the middle, connecting the private dining units of different sizes on the two sides. To lead the guests to pass through quickly without staying in the public area for too long, black wooden-texture materials are employed, wrapping the whole facade and connecting the separate units, creating a relatively oppressive and mysterious atmosphere. The white stone-texture floor, together with the lighting sequence, indicates the entrance clearly.

Walking through the frame-hidden door on the passage facade, guests are welcomed to a broader space where the excitement and confidence for a hidden environment is generated. The rice-paper texture screen is again adopted to decorate the window-side dividing interface to filter the natural light, making it possible for the tree shadow casted on the screen become the protagonist along with the dishes at a certain time.

The Waking
Learning techniques from the Song and Yuan Dynasties and absorbing spirits from the Jin and Tang Dynasties, modern people are looking for inspirations from the ancient times. The living style of the western people is straightforward and realistic. In terms of architecture, they believe that the functional needs must be the primary design concern. The rise of Modernism further enshrines the functional value of architecture so much that for a relatively long period of time, in the local Chinese cultural context, the " function " of the design was also hailed as the overarching concern. Actually, the " function " of the design is not that powerful as to be regarded as a cure-all medicine to solve the conflicts between humans and the artificial things. A comfortable space which offers real-time experience is the product of multi-dimensional information delivering. The so called spatial interpretation is led by the spatial awareness which is full of the adventurer's spirit. People, like the adventurer, aspire to climb to a higher height to see the grander view, while also prepare themselves to cope with the bewilderment and confusion if they fall off the mountain and are burdened by the illogical conflict-solving process. The former leads to breakthroughs and surprises while the latter is a negative response which may turn the design into a low-quality production.

就餐区的临窗立面，开窗杂乱之际光的肆意，用宣纸材质裱糊出的屏风墙，将自然光线进行二次过滤、晕染出糯白的儒雅气氛

The window-side facade receives randomly-casted sun light when the window opens. The rice-paper texture screen is able to filter the natural light for a second time, creating refined and modest atmosphere.

为促进食客快速通过，避免公区停留。黑色木饰面材质
制作的轻薄界面由顶面包裹至立面，并将各单元相互串
联，呈现出相对压迫且神秘的气氛。白色石材地面配合
灯光序列，成为引导进入的显性指示

In order to promote diners to pass fast,and to avoid staying in public areas, thin interface
between black wood finishes is wrapped up to facade, and each unit is series together showing
relatively oppressive and mysterious atmosphere. White stone surface matching lighting
sequence becomes dominant indication that guides entry.

高尺度厅堂存在的意义，是为自然空间向人造空间的过渡做好准备

The high story-height hall is designed to prepare for the transition from natural space to man-made space.

黑色条形钢板制作的区间分割
面，被设计者称为"空间面纱"的
序列，在食客行走的过程中，视
线不断被放大，发现食物制作的
过程当是欣喜

The sequence of partitions made of black steel strips, called"spatial veil"by the designer, provides diners with the pleasure of gradually discovering the process of food preparation as they walk around with their eyes enlarged.

一个舒适并拥有即时体验的空间，
是多维度信息传递的结果

A comfortable space offering
instant experience is created
through multi-dimensional
information delivery.

双层挑高尺度的错落关系，是体察
原有建筑内部空间后的理想塑造

The relationship of dislocati
between double layer high c
is the ideal model after obse
original interior space.

层高十余米的单层厂房内，脱离原建筑外墙，建立新的空间
秩序。回应相对市井面貌的外部环境，有必要不动声色地做
出气质转换

In the single storey workshop with a floor height of more than 10 meters,a new space order is established by separating the original building exterior wall. Responding to external environment that relatively has marketplace appearance, it is necessary to make a temperamental transformation quietly.

丙申年初，吴先生来访，欲造店城南为旗舰，取民国风韵谓之精神，名曰民国红公馆。选址停当，有南北二楼架桥相连，流水其间亭院各处，后置松柏、芭蕉、紫竹、桃花、杨柳等。

At the beginning of 2016, Mr. Wu came to visit me and expressed his idea of building a flagship restaurant in the old Nanjing area featuring in the Republican architecture style with the name of Republican Hong Mansion Restaurant. The location is well chosen, with a bridge connecting the two buildings in the north and the south separately, the water flowing within the pavilions, and pine trees, banana trees, black bamboos, peach trees as well as willows planted at the backyard.

红公馆造店记之烟雨江南

陈设 & 执行	蜜麒麟陈设组
灯 光 顾 问	Dark Light Lighting Design
艺 术 顾 问	乐泉
项 目 地 点	南京剪子巷
项 目 面 积	1 700 平方米
竣 工 日 期	2017.6
主 要 材 料	木 泥灰 水磨石 大理石
摄 影 师	李国民

Misty Rain of Jiangnan Made by the Hong Mansion

Soft decoration : Miqilin Display Group

Lighting advisor: Dark Light Lighting Design

Artistic advisor: Le Quan

Project location: Jianzi Alley, Nanjing, China

Project area: 1 700m²

Completion date: 2017.6

Primary materials: wood, plaster, terrazzo, marble

Photographer: Li Guomin

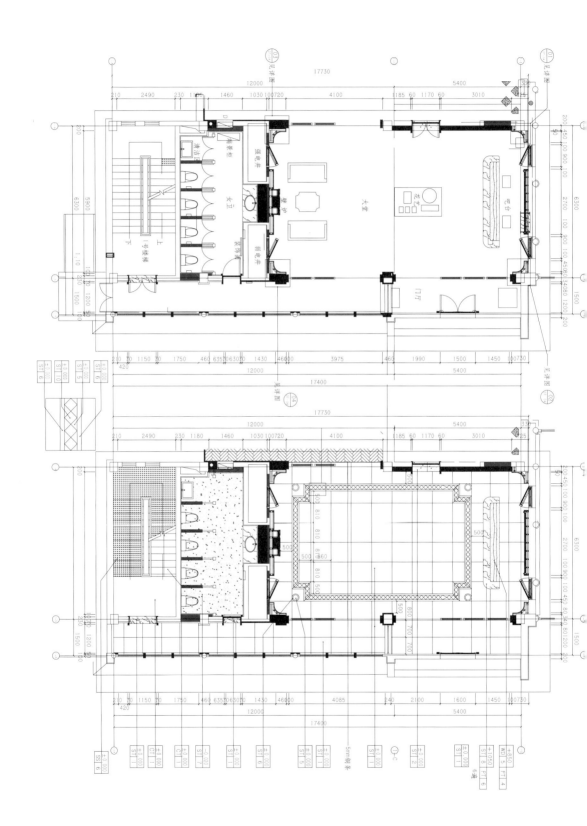

〔场〕

经老门东牌坊入剪子巷东三十米处，由北拾级而上，见五尺宽铜门向内，迎面玄关，屏风半掩，于转折处入公馆客厅。沿东西轴线布置吧台及礼宾区，中置岛台书榻，分离出内部交通，西侧吧台背景嵌入由艺术家独立创作的"南京故事"题材浮雕，辟邪、街巷、祥云、秦淮胜境等元素跃然画面，拉开通向民国往事之序幕。浮雕上方正中悬挂有被誉为"当代书峰"的乐泉先生创作的草书匾额"红公馆"，取材民国时期保存至今的紫桐木整板雕刻。吧台面放置磁石电话与上方灯盏呼应，书榻上布置百科旧籍、鸟笼、陶罐、烛台等细节，重温历史生活中最细腻部分，也许，公馆的主人正是如此书香世家。客厅东面礼宾区以壁炉为中心，墙面悬挂总统府旧照油画，与西墙面浮雕遥相辉映，粉彩绣墩与提花地毯一副娇滴滴的模样，优雅中透着仪式感。

客厅最为重要的作用除了迎宾送客亦是集散枢纽，南面并置两个入口，分别通往一层堂食区和北二楼的包间区。北二楼布置九个包间，取名"大千食园""逸仙别院""美龄客厅"等，分别以民国历史名人为线索取名，包间内布置除基本就餐功能所用的家具外，均以各人物性格展开叙事，还原记忆的画面。通往二楼的楼梯始终暗淡，甚至晦涩，不禁回忆起旧作《竹里馆》中对"通过空间"保有的情感"试图在有温度的交互中保持部分冷静，从而在步入另一个场域前，整理出独立的情绪"。而如今，"独立的情绪"只有挂在灰色墙面正中的那幅画与空间和解，画面中推开的窗扇伸向街巷，窗台下粉色的荷花感染着盛夏的余晖，信札刚写了一半便要邮寄出去？似是一个女人的波涛暗涌。正是，灰暗的梯段正是为了波涛暗涌！

堂食区分为东西二厅，由过廊相连，过廊保持"通过空间"一贯的营造态度——在黑暗中获取光明，向光性是通过空间具备隐晦体验感的保证。堂食西厅由原始建筑院落改建而成，保留院落中的主要树木，通过重组微观庭院形成区域视觉中心。加建部分用反支撑结构将楼板剥离原庭院地面，使之形成更加轻盈的建筑体量，宛若将现代装置置于古典庭院中。顶面采用双层透明采光顶结构，便于过滤光线与减少能耗，日光下顶面可以获得饱满且充盈的自然光线，夜晚由外部投射照明，光影层层重叠，雨天时可观察到顶面充斥着落水涟漪的视觉奇观。

南楼呈传统建筑形式中"对照厅"布局，改造后仅留东过道为室内交通贯穿南北，其余空间皆遵从建筑梁柱关系，分割为四个独立就餐空间，围天井于内，并以天井平面尺度退让至柱基位置，改造成一池静水飘然屋外，置风灯于水面，与室内交相辉映，行人通过、客人入座，皆可体验建筑落于水中的轻盈通透，消减了传统建筑室内相对压迫沉闷的感受。

由北拾级而上，见五尺宽铜门向内，迎面玄关，屏风半掩

Take steps upwards from the north, only to find a five-foot wide bronze door facing inwards, with half covered screen standing in the entryway.

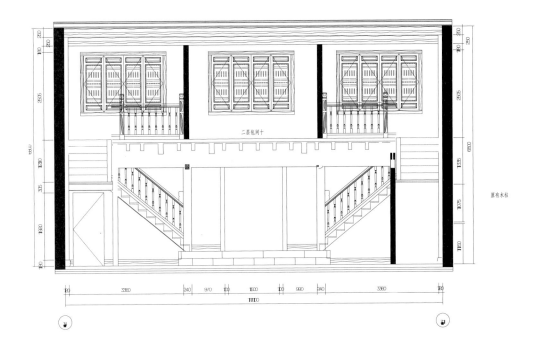

〔造〕

　　每一次营造都是温故知新。由意淫到修正，一个时代的营造手法应该尊重这个时代的技术纲领，并用艺术的方式呈现出来。空间产生于秩序限定，细节产生于逻辑细分。一个被工匠误解的细部做法有可能超越冥思苦想的细部创造，这是有可能的，并多次出现在现实之中。所以，营造亦需"破执"。一个构筑预想就像自然界中一个完整的生长，需要解读并抽离出最基本的基因，构筑物存在于构筑之前，一切必浑然天成。

〔性〕

　　民国空间的具象形态看似是笨拙的，未被细化的，却恰因此透出不同于其他时代的气质。如果你想获取一种光线，是弹性的或忧郁的，又或者，你希望创造一个轻盈的体量？如此种种，都只是当代解析之后的意淫，并产生于现代主义进入中国营造体系之后。用现代材料去表现之前的某个时代面貌或生活方式，像是用白话文解释古代诗词，唯有描述情境可以通达。万物起源、生长、变迁，外部力量皆不相同，生而有性，性生气息。

　　无缘生长于那个时代，对民国的理解，唯有感知，从书本史料里揣摩先人智慧，时而浮想联翩。邻家留声机的浅唱，厢房里的桌牌声，船坊琵琶声夹杂着嬉笑，街角转弯处着旗袍的女人。每一帧画面传递的即视感都是时空的烙印。而民国餐厅里就餐的人们应该是怎样的姿态？优雅的，闲适中透着一丝讲究。他们热爱自然，享受午后阳光，闲时修花弄草；他们偶尔谈论时代的焦虑，却不忘享受餐桌上的美好。

The Place

Walking east through the memorial archway of Laomendong to the Jianzi Alley for about 30 meters, then turning to the north and climbing the stairs, one walks though the wide-open bronze door, pass the hallway with the folding screen and enters the saloon of the mansion. The bar area and the concierge area are arranged along the east-west axis, with the island-shape book shelf arranged indicating the inner space. The background wall of the bar to the west is decorated with the theme embossment of "the Nanjing Story", which is the independent design of an artist. Elements such as Bixie (an auspicious beast in Chinese legend), alleys, clouds, and beautiful scenery of the Qinhuai area are all presented here, ushering the guests to the Republican period of China. Above the embossment there hangs the inscribed board which writes "Hong Mansion Restaurant" and is made of paulownia wood reserved from the Republican period. There is a magneto-telephone to echo with the light. The book shelf is decorated with the old encyclopaedia, the birdcage, the pottery jars and the candle holders, reviving the exquisite life details in the history. Perhaps, the owner of this mansion is from a literary family like this. The saloon area has the fireplace as the center, with the old photo of the Presidential Palace hanging on the wall, adding radiance to the western embossment wall. The pastel stools and jacquard carpet are elegant and delicate, reflecting a ceremonial grace.

The parlor serves not only as a place for guest reception, but also as a junction spot to other functional areas. There are two entrances to the south, leading separately to the dining hall on the first floor and the private box area on the northern second floor. There are altogether nine private boxes whose names are related to famous figures in the Republican period, such as "Daqian Dining Room" "Yixian Yard" "Meiling Parlor". The boxes are arranged not only to cover the basic dining needs, but also to tell a story of the figure and draw people back to the history. The stairs leading to the second floor appear to be dim and obscure, which reminds me of the design of my past work "Bamboo's Eatery". During the designing of the transition area of "Bamboo's Eatery", I stuck to the principle of trying to retain partial calm in the overall interaction so that an independent sentiment could be sorted out before entering another functional space.

每一帧画面传递的即
视感都是时空的烙印

Each frame conveys a sense of
deja-vu that is imprinted by
time and space.

Now, the independent sentiment is embodied in the painting hanging on the grey wall, compromising with the space. In the painting, the window is left open, facing the alley. Below the window there are pink lotus flowers reflecting the summer sunset. A woman seems ready to mail a half-written letter, delivering intense hidden emotions. Yes, the dim stairs are for intense emotions!

The dining hall is divided into the west hall and the east hall connected by a hallway designed under the principle of "seeking the light in the dark" for the transition area. The tendency to seek the light guarantees the dim atmosphere of the transition area. The west hall is rebuilt on the basis of the original yard with main trees reserved and the micro yard rearranged to form a visual center of the whole area. The add-on architecture adopts the reverse—support structure, lifting the floor surface above the yard ground, making it appear to be lighter and more flexible, as if the modern apparatus is implanted within the classical yard context. The ceiling is designed with double-layer transparent glass roof to filter the sunlight and save the energy. In the daytime, abundant sunlight provides the lighting, while during the night, the external lighting is arranged, casting overlapping shadows and lights. In rainy days, with the rain falling on the glass roof, ripples are generated which creates another fascination.

The arrangement of the southern building conforms to the layout of the traditional Duizhao Hall. After reconstruction, only the east hallway is kept to connect the southern and the northern space. The other areas are divided into four separate dining space by beams and pillars, with the courtyard designed within. A pool is built on the courtyard spot, adding beauty to the indoor space. Whenever there are passengers passing by, or guests coming in, they can appreciate the lightness and clearance of the design of locating the building on the water, which also decreases the relatively oppressive and dull sentiments that the traditional interior design brings.

The Design
Every design experience is the process to review the past and to gain new insights. From mere imagination to

modification process, we should respect the technological advancement and present it in an artistic way. The space is created out of order and confinement, and the details emerge from the logical segmentation. It is totally possible that a detailed handling which is generated from the incorrect interpretation of the craftsmen may turn out to be better than the original thoughtful design—it is proved in real life. Therefore, in designing, we sometimes need to abandon our obsessiveness. The design process is like the complete growth cycle of natural things. We need to interpret and exact the most fundamental elements and conceive the architectural design before the real product comes into being. Everything needs to be natural.

The Disposition

The concrete form of the Republican style architecture appears to be clumsy and sketchy, which exactly accounts for its uniqueness different from other periods in history. What is the design purpose? To create the light to be elastic or somber? Or to create a lissome appearance? All these are still thoughts under modern analysis and are generated when the modernism is applied in the Chinese construction system. To present some certain historical condition or lifestyle with modern materials is similar to explaining the ancient Chinese poetry with vernacular words and only by understanding the context can we achieve the final reconciliation. The origin, the growth, the change and the external power of all things

are different and all things are born with a disposition which leads to liveliness.

I have no chance to be born in and to live in that period of history. Thus my understanding toward the Republican period comes from the appreciation of historical materials, books and my imagination. Sometimes I hear obscure sounds from the neighbor's phonograph, from the shuffling of cards in the entertainment room, from the ship where people are laughing and the performer playing the Pipa. Sometimes I see women dressed in cheongsam at the corner. Every scene is as real as if I am at the spot which delivers distinct features from that particular period of history. How do people in the Republican period behaved at table? Were they elegant, comfortable or a little picky? They loved the nature, enjoyed the afternoon sunlight, and spent spare time in gardening. They talked about the anxieties of their time sometimes but still appreciated the beauty on the dining table.

四个独立就餐空间，围天井于内，并以天井平面尺度退让至柱基位置。改造成一池静水飘然屋外。置风灯于水面，与室内交相辉映，行人通过、客人入座，皆可体验建筑落于水中的轻盈通透，消减了传统建筑室内相对压迫沉闷的感受

Four separate dining spaces enclose the atrium, from which to the base of the columns a pond of still water is designed, with storm lanterns placed on the water surface, reflecting the interior. Pedestrians passing by and guests taking seats would experience the lightness and transparency of the building reflected in water, alleviating the oppressive and dull feeling indoor spaces traditionally generate.

南北二楼架桥相连，流水其间亭院各处，后置松柏、芭蕉、紫竹、桃花、杨柳

The north and south towers are connected by a bridge, with pavilions and courtyards surrounded by flowing water, and pines and cypresses, bananas, black bamboos, peach blossoms and willows growing in the back.

堂食西厅由原始建筑院落改建而成，保留院落中的主要树木，通过
重组微观庭院形成区域视觉中心

Converted from the original building courtyard, the west hall of the
canteen retains the main trees in the courtyard and creates the visual center
of the area by reorganizing the microscopic courtyard.

一个被工匠误解的细部做法有可能超越冥思苦想的细部创造

A misunderstood detailing practice by a craftsman could potentially
outshine the meditative creation.

加建部分用反支撑结构将楼板剥离原庭院地面，使之形成更加轻盈的建筑体量，宛若将现代装置置于古典庭院中。
顶面采用双层透明采光顶结构，便于过滤光线与减少能耗，日光下顶面可以获得饱满且充盈的自然光线，夜晚由外
部投射照明，光影层层重叠，雨天时可观察到顶面充斥着落水涟漪的视觉奇观

The floor slab of the extension is stripped off from the original courtyard ground with counter-support structure to form a lighter building volume, just like placing a contemporary piece of equipment in a classical courtyard. The roof adopts double-layered transparent structure, facilitating the filtering of light and reducing energy consumption, so that it can get full and abundant natural light in the daytime, and is illuminated by external projection, with overlapping layers of light and shadow at night. On rainy days, the visual spectacle of falling water ripples can be observed on the roof.

北二楼布置九个包间，取名"大千食园""逸仙别院""美龄客厅"等，分别以民国历史名人为线索取名

Nine private boxes are set up on the second floor of the north building, which are named after famous people in the history of the Republic of China, such as "Daqian Dining Rom", " Yixian Yard" and "Meiling Parlor".

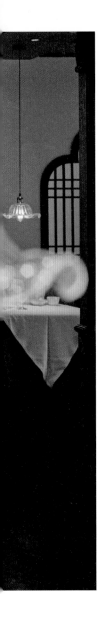

Singing on the boat swings back in red walls and flaps the green water from river banks. Time flies, dynasties vanish. A feeling of melancholy that as thick as juice is brewing from the hustle and bustle of life. A hundred meters to the west of the main hall of Confucius Temple, the gate of 17 street is opening for welcoming guests. It is said that it was a big world song and dance hall in the Republic of China era, crowded with literati. Nowadays, the restaurant of Red Mansion is built up here. The green and red light of the reflection on the river switches with the illusion of the eight beautiful ladies from Qinhuai River. Under the intuition impression given by the field, I reflexively think of Liu Rushi. I only recognize her as a patriotic poet, once dressed herself male and self-named "Liu Rushi" (Rushi here means confucian). Though she is only a woman in traditional society, but she has a deep emotion of "Family-country". When Ming Dynasty died, she attempted to give her life for the country carrying her husband. When her husband surrendered to Qing Dynasty, she tried her best to stop him. She also supported Rebel , from which can tell her ethnic integrity visibly. She interacted with celebrities, talking about prosperity and decline of the world. Liu Rushi once said, " Our country is now in trouble. It is the time that needs heroes to squash the revolt and resist foreign aggression. They should map out the strategy and repel the enemy like Xie Dongashan, and should not to be as bright as Tao Yuanming . If I were a man, I must save my nation from subjugation and ensure its survival, as well as devote myself to my country". At the end, in order to protect her dead husband's industry, she ended up killing herself to scare away the villains. That's when the talented woman ended her life. Among the eight beauties, Liu Rushi is the woman with the most noble and elegant character. She has noble and unsullied moral integrity, and her beauty is free from vulgarity. Even five hundred years later, women go in and out of the luxury family, when fixing their makeup, still smear the rouge of Liu Rushi.

船歌荡回在红墙，拍打河岸的一绿波水，六朝烟雨十里秦淮，在冗长的生活杂役中煲出浓浓如浆汁般的绵绸。夫子庙正殿以西百米处，面北拾级而上，贡院街 17 号的大门不知从何时开始敞开迎客，据说民国时为大世界歌舞厅，骚客繁华。如今红公馆中餐厅造店于此，桨声灯影处的一抹绿红与八艳佳话的幻象来回切换。场域给予的直觉印象，条件反射地想到柳如是，我只认她是个爱国诗人，曾以男妆相自名"柳儒士"，传统社会一介女流，却有着深厚的家国情怀。明亡之时，携夫殉国未遂，夫降清廷，极力劝辞，资助义军，可见民族气节。与名士往来，纵论天下兴衰，柳如是曾说，"中原鼎沸，正需大英雄出而戡乱御侮，应如谢东山运筹却敌，不可如陶渊明亮节高风。如我身为男子，必当救亡图存，以身报国"。最后，为护亡夫产业，结项自尽吓走恶棍，一代才女也终结了一生。八艳之中，最有风骨的女子，志操高洁，美貌自当脱俗。五百年后的女子出入奢侈一族，再整妆容，或可涂抹一下柳如是的胭脂……

红公馆造店记之胭脂水粉

设 计 单 位　名谷设计机构
陈设＆执行　名谷设计机构蜜麒麟陈设组
特约艺术家　乐泉 乐晓菊
项 目 地 点　南京市夫子庙贡院街
项 目 面 积　2 200 平方米
竣 工 日 期　2018.7
主 要 材 料　木 涂料 水磨石
摄　　　影　李国民

Rouge Gouache Made by the Hong Mansion

Design instiution: Minggu Design Agency

Display design & execution: Minggu Design Agency, Miqilin Display Group

Special guest artist: Le Quan, Le Xiaoju

Project location: Gongyuan Street, Confucius Temple, Nanjing, China

Project area: 2 200m²

Completion date: 2018.7

Primary materials: wood, plaster, terrazzo

Photographer: Li Guomin

贡院街 17 号据说民国时为大世
界歌舞厅，骚客繁华

〔体征〕

以"探索"为主题展开平面布局，沿街面不足九米开间的两层店铺，规划两千余平方米无自然采光
空间，可谓清奇。一层布置门厅、民俗博物馆、堂食区三个由外而内的递进式单元；门厅引入传统建筑"藏
经阁"的概念，将咨客服务、休息等候、地域性格融为一体；空间按轴线对称布置；吧台以传统食盒的
形态为原型，植入灯光并艺术化处理直奔主题；背景以秦淮风光为题材创作的浅浮雕升华了地域环境的
特色；墙面转折处均采用切角设计，柔化界面过渡。

由钢板制作的拱形门廊进入民俗博物馆，像是通过古老城门进入由记忆包裹的深巷；门廊两侧分别
布置微型照相馆和账房，并横向展开老物件的铺陈展示，围绕着衣、食、住、行的主题，诉说着那个时
代的生活点滴。博物馆是通往堂食区和二楼用餐区的必经之路，希望通过的食客在一个满载回忆的空间
稍做逗留或整理，并与之产生进入古典语境的准备。

继续往里由拱门进入一层堂食区，就餐单元由顶面分割所引发的组团律动，带动着平面单元式组合，

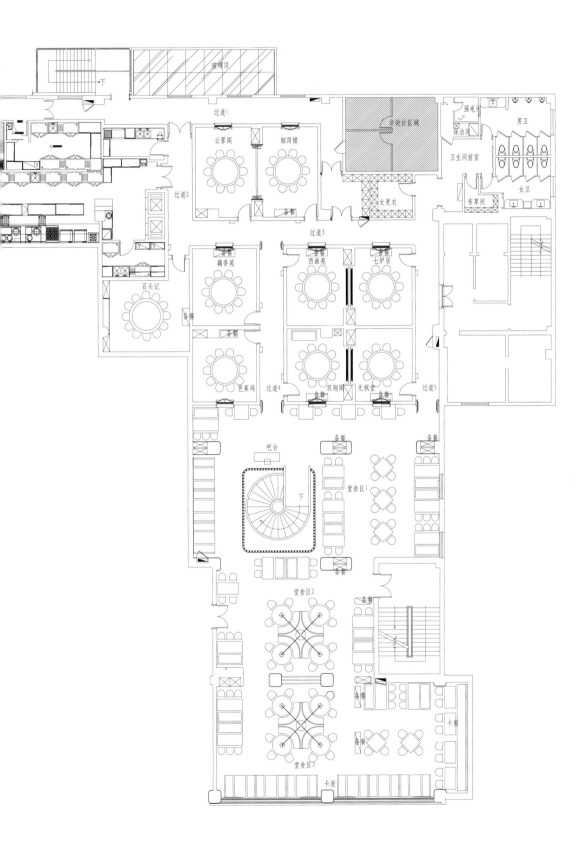

玻璃顶

过道1

云霁阁　烟雨楼

备餐

客设计区域

强电间

保洁间

男卫

卫生间前室

女卫

布草间

过道2

女更衣

过道3

石头记

藕香苑

备餐

西画苑　七炉居

备餐

备餐

备餐

芭蕉坞

过道4

双翅阁　无枫堂

备餐

过道5

备餐

吧台

下

备餐

堂食区1

备餐

备餐

下

备餐

堂食区2

备餐

备餐

卡座

堂食区3

卡座

并由在空中作横向延伸的灯具做出竖向呼应，单元顶面的阵列表现由黑色的缝隙连接，繁复而交叉的设备系统被整理消隐在缝隙之中。

通往二层的竖向交通由黑钢板与珊瑚红大理石制作的旋转楼梯连接，并放置在独立的空间内，逻辑设定为一楼到达二楼是由空间连接完成的，楼梯只是空间的一部分而已。于是，除了婀娜盘旋的楼梯完成基本输送功能，更重要的是需要行动的趣味，空间限定采用犹抱琵琶半遮面的竖向系统，用精工制作的铜质格栅围合出中庭的透视感，而中庭是需要轴心来建立心理依靠的，斗拱说："我才是轴心，不是谁都可以支撑中国性的力量部分！"于是以传统建筑构件斗拱为原型创作的艺术吊灯顺利占据了中庭的核心，与旋转的梯段一起成为行动的组织者。

进入二层，空间被划分为开敞式就餐与私密就餐单元。首先进入眼帘的，是由拱形门廊引发的连接

各包间的通道，包间入口的位置有意与对门做出避让，并以红楼梦、西厢记、桃花扇等文学戏曲名篇作为各包间的叙事主题，突显性格的同时，功能建设亦精彩。环绕墙面的线脚与主题画框线角均采用1厘米乘1厘米的实木阳角四十五度叠合做法，在不同的空间维度上做出阐述，或收边、或分割、或刻画细部，形饰语汇在不断的重复中建立记忆。

二层沿街面的开敞式就餐区本可以与街面的行人产生视觉互动，却因历史建筑立面保护的条件设立而变得遮蔽。整个店面唯一一处拥有自然采光的位置固然不能放弃建造的智慧，用古法玻璃与框架建立的视觉遮蔽界面，可以很好地让光线进入，同时屏蔽纷乱的外部形态。顶面有序的木质分割中局部嵌入浊银镜面，有效地拓展了空间纵深感，也可将街面动态部分输送到室内。

空间主体气氛以深灰色木饰面为主，精致的线角铜艺贯穿其中，搭配粉色系的主题表现。家具亦呼应空间，采用大量深色编织藤面的古典做法，搭配粉蓝色布艺相得益彰。厚重的历史长河与婉约瑰丽的秦淮妆容，在一个空间内完成了对偶的诗话。

〔行动的逻辑〕

生活美学的信众很难对逻辑表示认同，对空间的渴望亦是隐性的，除非来了个较真的建筑师，一边画着草图一边解释："一楼到达二楼是由空间连接完成的，楼梯只是空间的一部分而已。"还有对建设保有自信与经验的业主，面对低矮的建筑层高满脸沮丧。千秋大计毁于一旦的刹那，身着圣冠铠甲的建筑师踏云而来，诙谐地告诉他："有趣而生动的空间与层高没太大关系，空间张力才是制胜的法宝。"再有喜明者拒绝黑暗，喜暗者拒绝光明，他们只接受在文学里有伏笔，在佛法里有谦卑，在音乐里有律动，却很难享受建造过程的此起彼伏。空间的变化折射出动态逻辑，从不局限于先入为主的线索设定，因

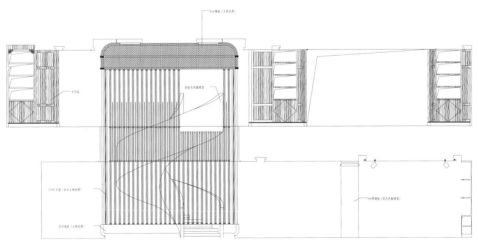

由钢板制作的拱形门廊进入民俗博物馆，像是通过古老城门的缩影进入由记忆包裹的深巷，门廊两侧分别布置微型照相馆和账房

The arched corridor made of steel plates leads you into the folk museum as if you were passing through the miniature gates of an old town into a deep alley wrapped in memories, flanked on both sides by a microphotography studio and an accountant's office.

家具呼应空间，采用大量深色编织藤
面的古典做法，搭配粉蓝色布艺相得
益彰，厚重的历史长河与婉约瑰丽的
秦淮妆容，在一个空间内完成了对偶
的诗话

As echoes of the space, the furniture uses a lot of classical dark woven
rattan surface, with perfectly complemented pink blue fabric; the
heavy long history and the graceful and magnificent Qinhuai scenery
compose a coupled poetry within the same space.

为什么所以什么，无形中陷入功能运算的深渊。就像在线索逻辑中一米九身高的人需要睡两米二的床，而在动态逻辑中根本不需要床。

〔流变的形饰〕

不知何时形饰作为表皮化的嘲讽，似是缺少内涵的表现主义，或多或少迎合了浮躁社会的价值核心，而偏离其解释美好的本质。现代主义传递出的异己力量，让作为个体的人感到无比的孤独，疏远与冷漠使后现代主义迅速地站出来，缓和了情感寄托的矛盾。柯布西耶在帕提农神庙的平面布局中得到现代主义建筑的启示，赖特从草原走向了光洁的流线，芝加哥学派将形式与功能设定为从属关系，在西方工业革命建立的新秩序与现代主义兴起后不断涌现的新观念中，粉饰的部分趋向消亡甚至批判。历史的镜头转向中华，2500年前的春秋鲁班团队、1800年前的魏晋竹林团队、1000年前的赵宋士大夫团队、500年前的朱明文人团队，似乎并不在意果走向哪里，却让建造、思想、生活、美学极尽繁华夺目的过程，那种以自我为中心，忽视公共空间生长的事实，在当下的社会语境中，更像是调节天平的砝码。当价值观念指向西方，曾经的文明缺陷亦可转化成新时代回归理性的基石，形饰再次登上中华后现代的舞台，成为生活的细节、情感的补充、建造的润滑剂。如果说古典之美在于骨骼健全的形饰，那么自信之美便让形饰有了重生的意义。

娴娜盘旋的楼梯完成基本输送功能、更重要的是需要行动的趣味

The graceful spiraling steps serve the basic transport function, and more importantly, convey the fun of mobility.

Signs and Symptoms

With "exploration" as the theme, the plane layout is carried out. The two-story shops with less than nine metres along the street are planned with more than 2 000 square meters of natural lighting space. It can be called strange.

The foyer, the folk museum, the dining area and three progressive units from outside to inside are arranged on the first floor. The foyer introduces the concept of "Classics Pavilion", which integrates customer service, rest and waiting area, regional character into one. The space is arranged symmetrically according to the axis. The bar takes the form of the traditional food box as the prototype, implanting lights and artistic treatment to go straight to the theme. On the background of Qinhuai scenery as the theme, the bas relief sublimates the characteristics of the regional environment. Shear angle design was adopted at the turning point of the wall, which softens the interface transition.

用精工制作的铜质格栅。围合出 中庭的透视感

The finely crafted copper grille strikingly conveys the sense of transparency of the atrium.

以传统建筑构件斗拱为原型创作的艺术吊灯。顺利占据了中庭的核心、与旋转的楼梯一起成为行动的组织者

An artistic chandelier, based on the traditional bracket system of buildings, occupies the center of the atrium, becoming the organizer of movements together with the spiraling steps.

Entering the folk museum through the arched porch made of steel plate enters folk museum is similar to entering alleyway wrapped by memory through an ancient gate. There are micro photo studio and account room on both sides of the porch separately, and horizontally expand old objects layout. It focused on clothing, food, shelter, transportation, telling the life of that era. Museum is the only way which must be passed to go to dining areas. Diners are expected to stay or organize themselves in a space with memories for a while, and prepare themselves to enter classical contexts.

Entering the dining area through the arch, the dining unit is driven by the group rhythm caused by the division of the top surface. The horizontal unit combination is driven, and the extended vertical lamps echoes in the air horizontally. The array expression at the top of the unit is connected by the black gap, with the complicated and crossed equipment system arranged and hidden in the gap.

The vertical traffic to the second floor is connected by a rotating staircase made of black steel plate and coral red marble. It is placed in a separate space. The logic is that the first floor to the second floor is connected by the space, and the staircase is only part of the space. Therefore, in addition to the graceful hovering ladder to complete the basic transportation function, the interest of action is more important. The space is limited by a vertical system and still holds partly concealed. The copper grille made by precision work violates the perspective of the "atrium". However, the "atrium" needs the "axis" to establish psychological support. Dougong said, "I am the axis, not everyone can support the Chinese power part!" So the art chandelier, which is based on the traditional architectural component Dougong, smoothly occupies the core of the atrium and becomes the organizer of the action together with the rotating staircase.

Entering the second floor, the space is divided into open dining and private dining units. The first thing that comes into view is the passage connecting each private room caused by the arched porch. The entrance of the private room intentionally avoids the door and takes the famous literary operas such as "Dream of Red Mansions" " Story of the Western Chamber" and the "Peach Blossom Fan" as the narrative themes of each private room. While highlighting the character, the functional construction is also wonderful. The line feet around the wall and the corner of the theme frame are all adopt the overlap method of 1 centimeter multiplied by 1 centimeter and 45 degrees of the positive angle of solid wood, which is expounded in different spatial dimensions, either closing the edge, or dividing, or depicting the details. And the shape vocabulary is built up in the constant repetition of memory.

The open dining area on the second floor along the street, which could have visual interaction with pedestrians on the street, has become obscured by the establishment of the protection conditions of the historic building facade. It is the only place in the store with natural lighting. The visual shielding interface is built by ancient glass and frame, allowing light to enter smoothly and shielding the chaotic external form at the same time. Within the orderly wood segmentation on the top, the turbid silver mirror is partially embedded, which not only expands the spatial depth effectively, but also transports the dynamic part of the street surface to the interior.

The main atmosphere of the space is mainly dark gray wood decoration, with exquisite copper wire angle through it. It matches the pink theme performance. The furniture also echoes the space, using a large number of classical methods of dark weaving rattan surface. Pink and blue fabric art complement with each other. Thick history and graceful Qinhuai makeup complete the dual poetry in the same space.

The logic of action
It is very difficult for believers of life aesthetics to agree with logic, and the desire for space is implicit. Unless, a serious architect comes to draw a sketch and explain, "The first floor to the second floor is completed by the connection of space,and the stairs are just part of the space" . Besides, the owners who have confidence and experience for their construction, depressed when facing the low building floor high. At the moment when the great plan of thousand years was destroyed in one fell swoop, the architect stepped on the cloud in the

holy crown armor, told the owners humorously, "Interesting and vivid space has nothing to do with floor height, and space tension is the magic weapon to win". Someone likes light but rejects darkness, and someone likes darkness but rejects the light. They only accept that there is "foreshadowing" in literature, "humility" in Buddhism, and "rhythm" in music, but find it difficult to enjoy the rise and fall of the construction process. The change of space reflects the dynamic logic, not limited to the preconceived clue setting, just like if there is a cause, there must be a result. It falls into the abyss of functional operation. Just like in the setting of "clue logic", people who 1.9 meters tall need to sleep in a 2.2-meter bed. But, in "dynamic logic", they don't need a bed at all.

Rheological decoration

I don't know when "decoration" was used as a superficial mockery, which seemed to be the lack of connotation of expressionism. It caters to the value core of the impetuous society more or less, but deviates from its interpretation of the beautiful essence. "Modernism" conveys the force of alienation and let the people as individuals feel extremely lonely. Alienation and indifference make "post-modernism" stand out quickly, and ease the contradiction of emotional sustenance. Corbusier was inspired by modernist architecture in the plane layout of the Parthenon. Wright moved from "prairie" to "smooth streamline", and the Chicago School set form and function as a subordinate relationship. The Luban team 2 500 years ago during the Spring and Autumn period, the bamboo forest team 1 800 years ago durning Wei and Jin Dynasties, the scholar bureaucrat team 1 000 years ago during Song dynasty and the literati team 500 years ago during Ming Dynasty did not seem to

care where the result goes, but let the eye-catching process of construction, thought, life and aesthetics extremely prosperous. Those kind of self-centeredness and the neglect of the growth of public space are more like the balance weights in the current social context. When the values point to the West, the defects of civilization can also be transformed into the cornerstone that returning to rationality in the new era. The decoration will once again step onto the stage of Chinese post-modernism and become the details of life, the supplement of emotion and the lubricant of construction.

If the classical beauty lies in the shape of sound bones, then the beauty of self-confidence makes the form have the meaning of rebirth.

环绕墙面的线脚与主题画框线角，均采用 1 厘米乘 1 厘米的实木阳角四十五度叠合做法，在不同的空间维度上做出阐述

The line feet around the wall and the corner of the theme frame all adopt the overlap method of 1 centimeter multiplied by 1 centimeter and 45 degrees of the positive angle of solid wood, which is expounded in different spatial dimensions.

整个店面唯一一处拥有自然采光的位置，用古法玻璃与框
架建立的视觉遮蔽界面，可以很好地让光线进入，同时屏
蔽纷乱的外部形态。顶面有序的木质分割中，局部嵌入浊
银镜面、有效地拓展了空间纵深感

The entire storefront has natural lighting location, using visual cover interface built by
traditional glass and frame. It can make light smoothly enter, while shielding chaotic
external morphology. Within the orderly wood segmentation on top surface, silver mirror
is locally embedded, which effectively expands spatial depth sense.

涌动需要包裹，才不至风崩离析。就像建造性格跳跃的空间需切分出组团并预设平衡，组团平衡才是理解非常态空间背后稳如基石的建造逻辑。空间创作需要在建立骨骼的同时，完成语汇传递，塑造品牌性格。许是没有多余的银子装点新衣，空间不容掩饰，但求比例气质。如同塑造美男子，需在生长初始阶段，给予健康充足的养分，包括除食物与物理环境之外的生存意志，其意志高贵，骨骼高贵。

武侠小说里常有幼年孩童，离开常人的生存环境，流落山野濒临绝境，再逢巧合奇遇，练成绝世武功，终成一代名侠。这种不断涌现的"奇遇"是支撑与推动整个故事发展的兴奋剂。英雄的每一次修炼都来自现实世界的偶发与突变，殊不知"偶发"只是大千世界的变幻熔炉中未曾谋面的那个结果而已。如同科学实验之前提出的假设，最终被证明，被认定为客观世界。

项目名称	My Hotel
软装设计	蜜麒麟陈设组
项目地点	南京市奥体大街
项目面积	600 平方米
竣工日期	2018.12
主要材料	氟碳钢板 大理石 木 涂料
摄 影	李国民 金啸文

Reshaping the Aesthetics of Composition
－ Tactile notes of My Hotel

Surging needs a package, so as not to fall apart. Just like building a space with a leaping character needs to cut into groups and presupposed balance. Only in this way can peopel understand the construction logic that is as steady as a cornerstone behind the abnormal space. Space creation needs to complete vocabulary transmission and shape brand character while building bones. Perhaps there is no extra money to decorate new clothes, so space can not be concealed, but the proportional temperament is required, just like shaping a beautiful man, who needs to be given healthy and adequate nutrients in the initial stage of growth, including the will to survive in addition to food and physical environment. The will of the beautiful man will be noble, so as bones.

Project name: My Hotel

Soft design: Miqilin Display Group

Project location: Olympic Street, Nanjing, China

Project area: 600m²

Completion date: 2018.12

Main materials: fluorocarbon steel plate, marble, wood, coating

Photographer: Li Guomin, Jin Xiaowen

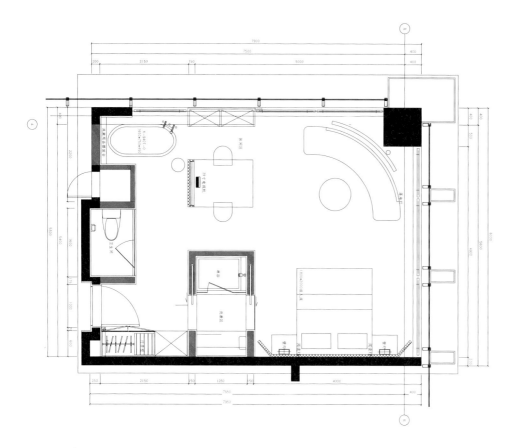

〔场所之舟〕

　　场所的伟大之处在于安放，无论大小、阴晴圆缺，总有一种最精准的方式将空间激活并大放异彩。建造者在其复杂的思想斗争与建造博弈中，无非只是在做一件事，那就是寻找，寻找那个通往人性的自然属性，赋予空间最基本的形态。建筑师于空间，在未知最终能呈现的空间面貌前，容易被理解为解决建造任务书提出的要求是设计的全部，并乐此不疲地表现出聪慧与灵巧，沉迷于技法的高超与丰富的物料，忽略直觉、画面感，允许局部破坏，一笔划过的飞白，留住从未停止的变化瞬间，让建造充满情感，趋向艺术性与生活的高尚感。

〔工法变迁〕

　　传统建造工法来自技工匠师，师傅带徒弟，多数是师者告知结果，照着葫芦先画瓢，画熟了缘由自悟。有好学者，站在师傅的肩膀上将工法演练纯化。正所谓不在工法的革命中消亡，就在工法的革命中崛起，即便是"崛起"充其量也就是"演化"，偶有创新者，我们不耐烦地集结称呼为"鲁班"。西方工业革命后的建造工法四通八达，建筑师时常参与科学家的运算，比如建筑模数的运算讨论，可考为现代主义建筑先驱柯布西耶与科学巨匠爱因斯坦于 1946 年在普林斯顿的会晤。新材料对于力学结构的帮助和高科技电脑运算的维度，使得工法脱离个体传承时代，走向无所不能的群体高科技时代。后现代主义打破了现代主义的僵局，企图让高科技变得有趣而生动，回望物料的情感与传统工法的自然。由东方智慧主导的建筑理念，将身体探向情感寄托的土地，希望触摸建筑的边界，尝试理解"无我"的生长感，工法游离于高科技运算与民间土木之间，无序、偶发、实验、对话自由切换，工法进入融汇再生长时代。

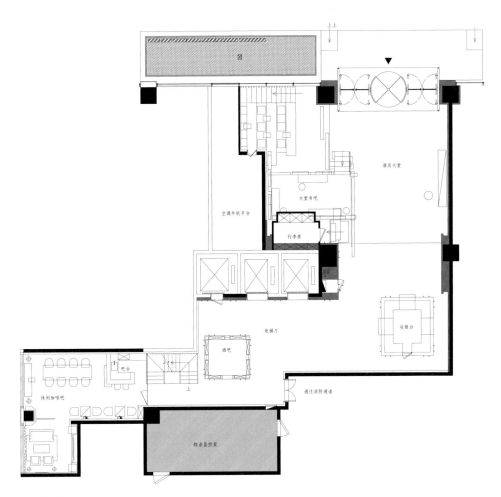

〔思索简报〕

　　行旅酒店，围绕短暂居住的基础性配置，抹去灯红酒绿的贴面，舍弃多余的商务配套，把住宿变得直接高效。公共空间的布局形式强调自由平面带来的律动体验，将入口、大堂吧、入住办理、电梯厅、酒吧分别放置在五个独立的盒子空间里，再排布出如同街边偶遇的位置特征。盒体与盒体之间的排布关系限定出必然的交通组织，其逻辑解释为由形体引发的动线布局，再将功能放置其内。设计者希望用非平台面化的构成形式营造出立体的新鲜感，来回馈投资人与住客对时尚的理解。

　　客房在平面布局上做出探索。为了让睡眠区的气氛变得舒缓而安全，削减客房卫生间的占地体量，采用半高的弧形墙面限定出需要遮蔽的私密空间，并与床头护墙板的末端弧线形成呼应关系，内部分割变得轻松而流畅。将盥洗、冲淋、马桶分离放置，或根据平面条件任意组合，如此，卫生间内部的交通空间被分享为客房内共有交通，使客房入口变得有趣。

　　连接酒店公区和客房的通过空间强调节奏的直接感，忽略平面铺陈，淡化入口形式，墙面深色的木饰面将空间压暗，灯光成为顾客进入的向导。在客房层每层的中心位置设立小书吧、艺术体验馆和小型健身房，缩短配套设施的服务距离。

Boat Of Place

The greatness of the place lies in its placement. No matter whether it is big or small, complete or incomplete, there is always a most accurate way to activate space and shine. In its complex ideological struggle and construction game, the designers are just doing one thing, that is, looking for the natural attribute that leads to human nature and giving space the most basic form. Architects for space, before the final appearance of the space is unknown, can easily be understood as solving all the requirements of the design put forward in the construction task book, and are happy to show intelligence and dexterity all the time. They indulge in the superb skills and rich materials, ignore intuition and the sense of imagery. They allow local destruction, like a stroke of flying white. They retain the moment of change that never stops, and make the construction full of emotion. They tend to the artistry and the noble sense of life.

The Change of Construction Method

The traditional construction method comes from craftsmen. Most of the results were informed by the teacher. Apprentices initially draw ladles according to the gourd, and the proficiency comes from self-study. There are studious apprentices who purify the construction method drills,standing on the shoulders of their masters. It is said that either perish in the revolution of the construction law, or rise in the revolution of the construction law. Even if there is "rise", that is just "evolution" at best. Occasionally, there are innovators. We impatiently assemble them and call them "Luban".Since Industrial Revolution in the West, there have been a variety of building models. Architects often participate in calculations with scientists, such as the calculation and discussion of building modules, which can be the reference like modernist architecture pioneer-Le Corbusier and the scientific giant-Einstein at the Princeton meeting in 1946. With the help of new materials to mechanical structure, and the dimension of high-tech computing, building models are isolated from the era of individual inheritance into an omnipotent era that full of high technology groups.Postmodernism broke the deadlock of modernism and attempted to make high-tech become interesting and vivid. Overlooking the emotion of materials and the nature of traditional methods, the architectural concept that dominated by oriental wisdom abandoned the sky. It reached to

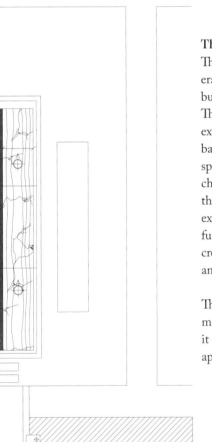

the earth, hoping to explore the boundary of architecture and trying to understand the growing feeling of "forgetting oneself". The construction method is separated from high-tech computing and civil engineering. Switching from disorder, accident, experiment, conversation freely, the construction method has entered the era of fusion and regeneration.

The Brief of Thought

The travel hotel, around the basic configuration of temporary shelter, erases the bright and colorful veneer, and gives up the superfluous business facilities and makes the accommodation direct and efficient. The layout form of the public space emphasizes the rhythmic experience brought by the free plane. It places the entrance, the lobby bar, the check-in hall, the elevator hall and the bar in five separate box space, and then arranges the location characteristics as if they met by chance on the side of the street. The arrangement relationship between the boxes limits the inevitable traffic organization, which is logically explained by the moving line layout caused by the shape. Then, the function is enclosed in it. Designer hopes to use non-planar forms to create a three-dimensional sense of freshness to give back to investors' and residents' understanding of fashion.

The guest room makes an exploration in the plane layout. In order to make the atmosphere of the sleeping area become smooth and safe, it reduces the covering volume of the bathroom in the guest room. It applies the half-high arc wall to limit the private space that needs to be sheltered which forms an echo relationship with the end arc of the bedside wallboard. The internal division becomes easy and smooth. The lavatory, flush and toilet are arranged separately, or combined arbitrarily according to flat conditions. So the traffic space inside the bathroom is shared as mutual traffic in the guest room, making the entrance of the guest room interesting.

The passing space connecting the hotel public area and the guest room emphasizes the direct sense of rhythm, ignoring the flat layout and diluting the entrance form. The dark wood finish of the wall darkens the space, and the light becomes the guide for customers to enter. A small book bar, an art experience hall and a small gym are set up in the center of each floor of the guest room floor to shorten the service range of supporting facilities.

艺术体验馆　　art experience gallery

公共空间的布局形式强调自由平面带来的律动体验

The layout of the public space emphasizes the rhythmic experience of free planes.

为了让睡眠区的气氛变得舒缓而安全，消减客房卫生间的占地体量，采用半高的弧形墙面限定出需要遮蔽的私密空间，并与床头护墙板的末端弧线形成呼应关系，内部分割变得轻松而流畅。将盥洗、冲淋、马桶分离放置，或根据平面条件任意组合，如此，卫生间内部的交通空间被分享为客房内共有交通，使客房入口变得有趣

In order to make the atmosphere of the sleeping area become smooth and safe, it reduces the covering volume of the bathroom in the guest room. It applies the half-high arc wall to limit the private space that needs to be sheltered which forms an echo relationship with the end arc of the bedside wallboard. The internal division becomes easy and smooth. The lavatory, flush and toilet are arranged separately, or combined arbitrarily according to flat conditions. So the traffic space inside the bathroom is shared as mutual traffic in the guest room, making the entrance of the guest room interesting.

光
之
叙
事

Story of Light

设计单位　南京东琪建筑
项目地点　南京市剪子巷
项目面积　1 500 平方米
主要材料　木 涂料 水磨石
竣工日期　2018.1
摄　　影　金啸文

Design team: Nanjing Dongqi Architecture
Project location: Jianzi Alley, Nanjing, China
Project area: 1 500m²
Primary materials: wood, paint, terrazzo
Completion date: 2018.1
Photographer: Jin Xiaowen

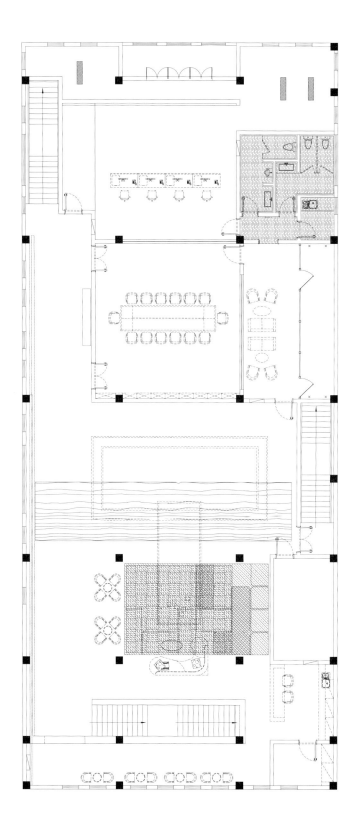

时间无垠　予筑新生　四十号

桌上放着一张照片，照片的下方打着"#40"的标记。我已记不清是哪一位同事完成了这张历史剪影的记录。只见面向街道的立面几乎完全笼罩在高大乔木的斑驳树影中，墙面上的马赛克材质骄傲地提醒着人们她那曾经辉煌的过去和引人注目的资历。

项目名称	四十号
设计团队	名谷设计机构建筑组
室内设计	名谷设计机构联合组
灯光顾问	Dark Light Lighting Design
项目地点	南京市来凤街
项目面积	3 600 平方米
主要材料	钢板 磐多魔 金属氟碳喷涂
竣工日期	2020.5
摄　影	夏至

Project name: #40

Design team: Architecture team, Minggu Design Agency

Display design: Joint team, Minggu Design Agency

Project location: Laifeng Street, Nanjing, China

Project area: 3 600m²

Main materials: steel plate, pandomo,
metal fluorocarbon coating

Completion date: 2020.5

Photographer: Xia Zhi

#40

A photograph stands on the table, with "#40" marked on the bottom. I can barely recall which colleague made the record of this historical silhouette. The facade facing the street is almost fully covered in the dappled shadows of tall trees. The mosaic texture on the wall is a proud reminder of her glorious past and striking experience.

将设计融化成历史的一部分，并试图把客户的性格特征和品位转化为建筑意向

Fuse the design into part of its history. We also attempt to translate
the client's personality and preferences into architectural intentions.

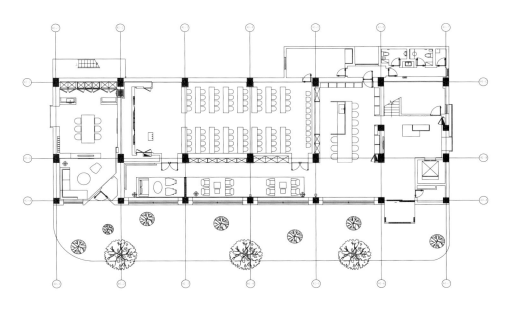

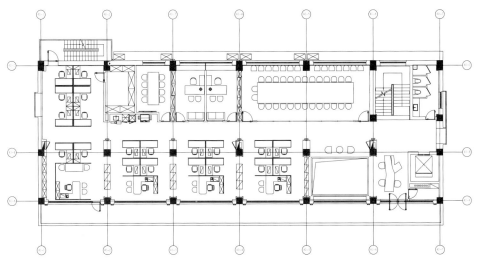

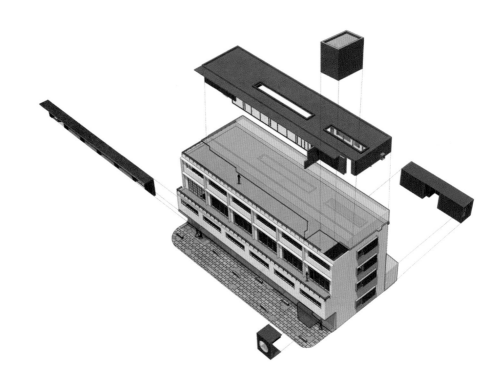

　　这是一栋老厂房，坐落于南京国家领军人才创业园内。园区历史悠久，向上可以追溯到清朝的江南银元制造总局，于民国时期相继更名为"财政部南京造币厂"及"度量衡造所部第二厂"，最终于1959年定名为"第二机床厂"。园区内的众多建筑就像一座座紧凑的岛屿紧密地构成一片群岛，很少会有合适的角度能够对四十号进行全方位的观察，于是得到了照片上所呈现的略显别扭的视角。

　　诚然，历史拥有价值，时间才是永恒的节奏。那些存在过的激情和热烈只能成为她所经历过的年代的标识。我们需要帮助她在当下塑造新的身份，找寻新的价值，在变化中打造出新的容貌。本次设计，团队要完成2 300平方米的一流办公空间，并将原建筑由4层加建至5层高度，进一步完善辅助服务体系以满足金基城市公司发展之必须，起到增强团队凝聚力、强化未来发展信心之作用。我们很清楚自己无法罔顾所处的历史环境。只关心个人创造的魅力，势必会贬低建构整体行为的作用。因此选择承认固有的空间性格并呼应她，创造出适宜的更新，将设计融化成历史的一部分，并试图把客户的性格特征和品味转化为建筑意向，揭示使用主体的部分特征，以非自我炫耀的方式代表某些内在价值。

　　设计保留了原建筑的质朴气质，传递出轻微的反都市化倾向。与土地和周边的联系是首层最重要的特征。利用透明质感，沿街一侧不再局限于空间的围合，反复强调着内部和外部的空间延续，相对公共和相对私密的空间在内外交织中和谐结合。于建筑外部精心组织建筑体验，诸多独特而富有层次的细部还增添了空间的戏剧性。外部场地利用人工形式的松散编织暗示辐射场域的边界，而入口处由内向外，建筑空间不断衍生，朦胧的东方叙事语境缓慢吟唱着孤独创造者的浪漫神秘感。

相对于底层的"虚"，建筑的上部保留了其"实"的特征。建筑体量不算巨大，但它营造的是一个特征强烈的内部世界，给人以深刻印象。方形的向心和稳定的力量传递成为建筑平面的主导特征。原建筑的长条体量及固有柱网结构确定了轴基线的方向。根据每层功能倾向的差异设置一条或多条垂直次轴线，进一步发展成组织骨架。办公空间组织的动机非常明确，并没有过多的强调秩序与等级，只是根据实际需求，简单设置功能位置，以求拉近环境亲密度。局部流线型的办公设置唤起漂浮感，个性化组合与普遍性概念的协调构成了这个方案的平面特征。垂直方向做着与平面均衡的努力。由于原有建筑结构不足以支撑大面积的中庭，于是本案选择了小型的跃层中庭，在墙体与空间中交替上升，以天空为领地，在顶部天光庭院完成释放。"微中庭"形式的确立，打破了单层平面的平淡布局，在局部体现出紧凑的向心意识，提炼出集体意识的建筑核心。同时作为功能及景观的定向标，暗示着公共性与私密性的边界。主要功能楼层 2F-4F，大面积的服务性区域结合交通空间展开，以确保实现交流的空间延续。位于 5 层的健身服务系统使这一切达到高潮，气质古典、色彩明快、肌理精致，谦虚地赞美材料的自然属性。这里气氛轻松，没有压迫，大家可以共享一个大型空间，分享一个共同时刻，这些都会给予使用者对场所的自豪感。

尺度、比例、材料、光的触感与自然界的联系，这一切使四十号成为令人期待的办公建筑。而我始终坚信，即使秋月草木凋零，竹影后的风景亦不孤单。

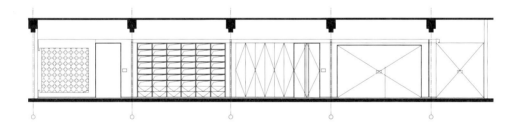

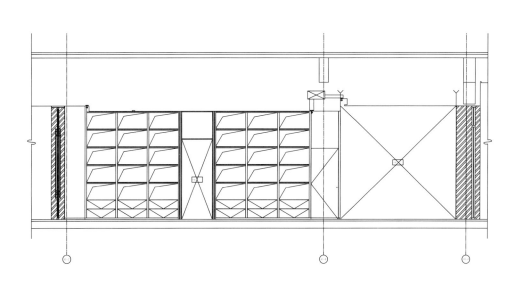

This is an old factory building sitting in the Nanjing National Leading Talents Pioneer Park. The long history the park can trace back to the Jiangnan General Mint of Silver Coins in the Qing Dynasty. It was then renamed as Nanjing Mint of the Ministry of Finance and the Second Factory of the Ministry of Weights and Measures successively during the Republic of China, and further as the Second Machine Tool Factory in 1959. The many buildings in the park are like a compact archipelago of islands, making it hardly possible to get a comprehensive view of #40 from any possible angle, thus resulting in such a slightly odd perspective in the photo.

It is true that history has value, but only time will sustain. The passion and enthusiasm of the building could only be a sign of the times she lived through. Our job is to shape a new identity, identify a new value and create a new face in the midst of change for her. In this case, the total area of design is 2 300 square meters of first-class office space, and one more floor is added to the original 4-storey building, so as to further improve its supporting services system, thus satisfying the development needs of the Kingjee Group, enhancing team cohesion, and strengthening confidence in future development. We are fully aware that we cannot ignore the historical context we live in. Focusing on individual creation only would inevitably devalue the role of the entire design and construction. Therefore, we choose to acknowledge the inherent character of the space and respond to it, and on this basis, make appropriate upgrading to fuse the design into part of its history. We also attempt to translate the client's personality and preferences into architectural intentions, and partially reveal the subject's characteristics and certain intrinsic values in a non-self-aggrandizing way.

The design retains the rustic qualities of the original building and conveys a slight anti-urban tendency. Connecting the land and its surroundings predominantly features the first floor. Using transparent textures, the street side is no longer confined to a spatial enclosure, but repeatedly emphasizes the spatial continuity between the interior and exterior, where public and private spaces are harmoniously woven together. The architectural experience starts to be carefully structured from the exterior of the building, with a variety of unique and layered details used to add to the dramatic properties of the space. The exterior site makes use of a loose weave of artificial forms to suggest the boundary of the radiating field, while at the entrance, the architectural space is constantly evolving from the inside out, with a hazy oriental narrative context slowly singing the romantic mystery of a solitary creator.

The upper part of the building retains its "concrete" character as opposed to the "virtual" sense of the ground floor. The building is not big in size, but it creates a strongly characterized interior world that leaves people strong impression. The centrality of squares and the stable transmission of power are the defining features of the building's plan. The long space and the inherent network of columns in the original building determine the direction of the axis. One or more vertical sub-axes are established according to the function orientations of each floor, and further developed into an organizational support structure. The office space has a fairly explicit and simple structure of functions based on actual needs to cultivate physical closeness, without too much emphasis on order and hierarchy. A partially streamlined office set-up evokes a sense of floating, and the harmony between individualized combinations and the notion of universality characterizes this plan. Vertically, an effort is made to balance the flat plane. Since the existing building structure is not strong enough to support a large atrium, a small and skip-floor atrium is adopted in this case, rising alternately through the walls and space, with the sky as the territory, and a skylit courtyard at the top to complete the function. The establishment of the "micro atrium" breaks the bland layout of a single-story plan, but reflects the compact centripetal consciousness through parts of the building and refines the core of the collective consciousness. At the same time, it serves as a directional marker of function and landscape, and suggests the boundary between the public and private space. The main functional floors from 2F to 4F are equipped with large service areas and traffic space to ensure the spatial continuity of communication. The fitness service system on 5F bring all this to a climax. The classical atmosphere, bright colors and delicate texture all modestly celebrate the natural properties of materials. In such a relaxing and unpressured atmosphere, people can share a large space and common moments, all of which will instill a sense of pride into space users.

The scale, proportions, materials, touches of light, and connection with nature, all these make #40 a desirable office building. I always believe that even if grass and trees wither in autumn, the landscape under bamboo shadows will never be lonely.

05

04

03

02

01

方形的向心和稳定的力量传递成为建筑平面的主导特征。原建筑的长条体量及固有柱网结构确定了轴基线的方向

The centrality of squares and the stable transmission of power are the defining features of the building's plan. The long space and the inherent network of columns in the original building determine the direction of the axis.

由于原有建筑结构不足以支撑大面积的中庭，于是本案选择了小型的跃层中庭，在墙体与空间中交替上升，以天空为领地，在顶部天光庭院完成释放

Since the existing building structure is not strong enough to support a large atrium, a small and skip-floor atrium is adopted in this case, rising alternately through the walls and space, with the sky as the territory, and a skylit courtyard at the top to complete the function.

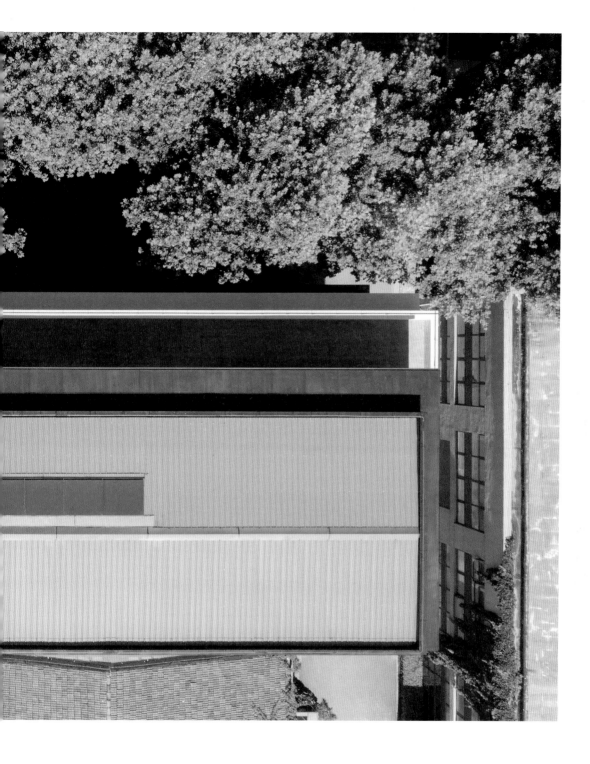

众多建筑就像一座座紧凑的岛屿紧密地构成一片群岛，很少会有合适的角度能够对四十号进行全方位的观察，于是得到了照片上所呈现的略显别扭的视角

The many buildings are like a compact archipelago of islands, making it hardly possible to get a comprehensive view of #40 from any possible angle, thus resulting in such a slightly odd perspective in the photo.

与土地和周边的联系是首层最重要的特征。利用透明质感，沿街一侧不再局限于空间的围合，反复强调着内部和外部的空间延续，相对公共和相对私密的空间在内外交织中和谐结合

Connecting the land and its surroundings predominantly features the first floor. Using transparent textures, the street side is no longer confined to a spatial enclosure, but repeatedly emphasizes the spatial continuity between the interior and exterior, where public and private spaces are harmoniously woven together.

入口处，由内向外，建筑空间不断衍生，朦胧的东
方叙事语境缓慢吟唱着孤独创造者的浪漫神秘感

At the entrance, the architectural space is constantly
derived from the inside out, with a hazy oriental
narrative context slowly singing the romantic mystery
of a solitary creator.

建筑体量不算巨大，但它营造的是一个特征强烈的内部世界

The building is not big in size, but it creates a strongly characterized interior world.

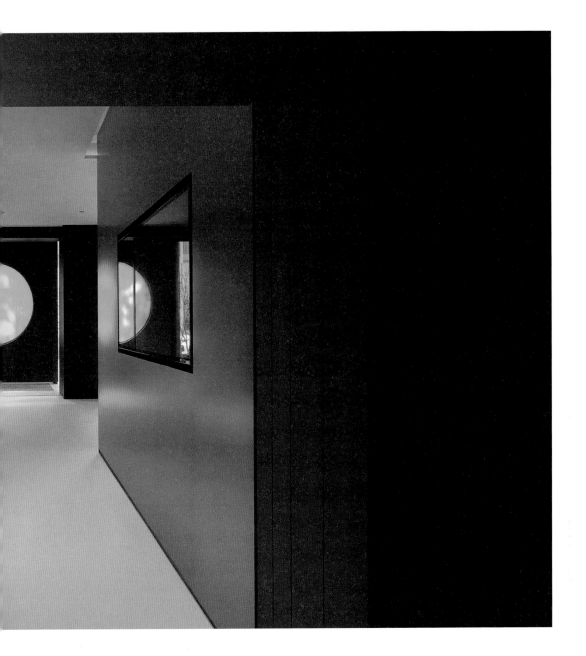

这里气氛轻松、没有压迫，大家可以共享一个大型空间，
分享一个共同时刻，这些都会给予使用者对场所的自豪感

In such a relaxing and unpressured atmosphere, people can share a large space and
common moments, all of which will instill a sense of pride into space users.

Q&A

建筑不是妥协，
设计更需要克制

《室内设计师》58 期访谈手记

<u>你如何理解什么是建筑？您的设计理念或者信条是什么？</u>

我理解的建筑，作为一种空间限定，负责沟通人与自然之间的关系，负责修复人类的生活秩序。人们常说文化是人类社会的精神与灵魂，那么建筑作为文化的重要载体，与文化互为表里，反映了其所在历史区间内的社会心态及客观要求。一旦成型便满载着所处社会时空的文化记忆，充当人与人之间的情感纽带。我们常说，人类是短暂的人类，自然是永恒的自然。建筑作为人类抵御天敌的工具，从诞生伊始就注定了它的反自然属性。因此我们在设计的时候更应该抱以谦虚的态度，摒弃哗众取宠的逻辑。"受自然恩惠，且求平等，不欲瞩目"这种不卑不亢的建筑性格为我所欣赏，在设计时相对会比较关注"克制"与"梳理"两个方面。美好的事物有很多，表达的方法有千万种，在一个项目中，前期考量时我会一直做减法处理，精炼出最想述说的内容，争取以最恰当的方式表达出来。

哪些建筑师、建筑作品对您的理念产生过影响？

建筑史上从来不缺少优秀的建筑师，瑞士的 Peter Zumthor（彼得·卒姆托），葡萄牙的 Alvaro Siza（阿尔瓦罗·西扎），还有芬兰现代主义建筑大师阿尔瓦·阿尔托等。其中有四位对我的影响最深，两位来自西方，分别是美国的路易斯·康和意大利的卡罗·斯卡帕。路易斯·康关注建筑秩序的恢复，以及自然与光线、自然与自然的关系解读。他呈现空间秩序的方式及理论系统对我的启发非常大。卡罗·斯卡帕在他的建筑范畴内，是难以被超越的。我曾经撰写过相关他的报道。前年有幸感受过他人生最后一个作品的现场，步入布朗家族墓园后让人感受到一种令人震撼、泪流满面的空间力量。可以说卡罗·斯卡帕先生在细节上的把控程度已经超出了建筑的界限。两位东方的设计师一位是杰弗里·巴瓦（Geoffrey Bawa），来自印度洋上的泪滴——斯里兰卡。杰弗里·巴瓦作为国宝级的建筑师，他的作品同属于现代主义建筑，但同时他建立了自己的东方认知，植根东方，做出了很多代表自身民族的作品。最后一位也是对我影响最深的建筑师是同济大学的冯继忠先生。冯老在一个特定的时期，一个物质和认知都缺乏的年代，独立思考，完成了方塔园何陋轩如此高水平的创作。就是这样一个以宋式美学为参照并为当代所认知的作品，虽经数十年风霜，今日仍熠熠生辉。冯老是我国不可多得的在建筑实践方面产生巨大影响的人物，其作品展现中国风骨，对我启发巨大，并以为明灯指引自己上下求索。

在您看来，您所毕业的学校以及在那里的职业训练对您现在的职业有哪些帮助？

我的母校是安徽建筑大学，在我上学的时候还是其前身安徽建筑工业学院。很惭愧，当时在校时并未专心于学习，更多的时间和心思都花在了搞音乐、组乐队上。现在回想，母校的教育体系其实非常的系统、完整、高效，专业针对性也很强。学校在进行教学的时候把很多不同专业的内容进行融会贯通，将环境艺术和建筑、规划、景观作为完整共同体去学习，仅在毕业前一两年根据专业不同而有所偏向，潜移默化地培养了学生完整的建筑视野。虽属环境艺术专业，仍要学习建筑学的基础课程。当时在校觉得学习理论任务繁重且枯燥无味，真正拿到项目又感觉无从下手，使不上劲。后进入行业，每年接触很多项目，逐步感觉到学习理论的重要性。很多基础课程比如画法几何、建筑初步等都体现出自身含金量，曾经受过的手绘训练、理论训练、思想训练、建筑训练都对职业发展起到了决定性的作用。

毕业、创业到陆续接项目，作为新锐建筑师／设计师，您觉得执业初期哪些经验值得分享？

毕业后走上社会，开始接触实际的项目运作，才渐渐明白了什么是设计。在校时对设计的理解仅限于学科，毕业后就要通过不断的实践去加深对专业的认知。我认为，一个设计师着手设计的初期必须是积累经验，积累知识，积累人与人的交往技能，这个不是在专业课堂上就能解决的问题，需要长时间的摸索和思考。积累的同时亦须坚持，要成为一个成熟的设计师必定遭遇到很多磨难；即便相对成熟，也会遭遇到职业的瓶颈、项目的阻力。每当这个时候都是考验职业信仰的时候，我们需要足够坚定，坚守着去等待机会，等待能够表达自己设计思想和方法论的途径。厚积薄发，待机会来临时又将是考验积累的时候。要想一个设计方案最后成为作品，中间付出也非常艰辛。从最初的创意到执行阶段的图纸、施工图，到解决现场矛盾，执行体系必须非常完整，需要面面俱到，需要事必躬亲，只有这样才能呈现出表达相对完整的设计意图。

回顾之前的作品，您觉得哪些作品在设计历程中具有一定的代表性，或是体现了您在那段时期内哪些思考？

回顾之前作品，其实是有递进式的在完成一个系统内的思考，逐步建立自己的认知。2013 年的时候，一个偶然的机会让我接触到传统的建筑，提供了在传统建筑内实践的可能。当时我认为传统与当代需要进行连接沟通，于是《小东园》创作过程中，我把关注点放在"对话"上。随着新的作品陆续展开，摸索深入，慢慢发现传统与当代并不单单是对话那么简单，这里有一个时空轴的概念，我们急需对传统的理解，先辈留下的文化瑰宝我们忘却得太多，传承得太少，甚至不具备读懂她的能力。为了学习理解认知传统的方法，我开始阅诗书、读画、看山水。常说心中有山水才是真正的山水，"山水"于中国文化代表着内在。在设计《印象村野》的时候，我开始尝试表达自己心中的山水。这个作品提炼了很多内心情感，使用抽象的手段传递意向，用抽象去表达具象，用抽象去传递内心。这同样体现在《桔子水晶》里。重重叠叠的造型构筑起整个设计，芦苇为灯，秋千为椅。自然界的事物在这个空间内，重新被理解，展现新的秩序。2014年的下半年，工作室得机缘在老门东历史街区落地。一个三进的传统院落，可以完全摒弃商业左右，为工作室自主。世间最美无非阳光、空气、雨露，当我们连这些都进行取舍，该是如何一种状态？第二进院子—"来院"承担了这种想法的空间实践。我们首先确定了光线的选择，摒弃室内照明道具，通过控制光线的路径及形状，引导自然光线进入室内。屏蔽了大部分色彩，通过降低空间饱和度等一系列手段，将空间清理出拙朴干净的气质。建筑解决功能问题和美学问题就足够么？她的本质到底是什么？仍是需要持续探索的课题。

您在设计过程中比较关注哪些方面？这些是否对项目最后的完成度有帮助？

每一个项目都有其特殊性，其本身的性格就是项目的灵魂所在，相应的每个项目应该有其特有的创新点，如何表达会成为项目把控的出发点。其次是执行问题，设计体系必须面面俱到、层层相扣。比如说空间表达、光感表达、软装表达，等等，凡是我们的手、眼、行为能接触到的地方，都必须递进式地执行细化。设计之初关注品格创新，设计阶段细致入微、善意体贴，最后呈现的作品一定是优秀的。

最近在忙哪些新项目或者研究？

最近思考的主要是生态能源方面的问题。改革开放以后大量的土地开发已经让国家和行业呈现出疲劳状态，造成程度巨大的环境污染。在此情境下，我开始探索更环保的物料，那些物料或者自然生成或者自动更新或者具备呼吸属性。近期着手的是一些关于竹子的实验，一个茶艺场所，一个人居住宅。尝试用一些很廉价的素材去构筑启发心灵思考的质朴设计。

在当下的设计现状中，您的事务所 / 工作室采取了哪些应对措施？

当下有两种现象比较突出，一种来自于经济环境，一种发自行业内部。在房地产经受调控后，设计行业也受到了一些限制。情况越复杂，我们越要更多地关注设计本身，提高核心竞争力，打造更优秀的作品，坚定不移地贯彻团队精神。近几年行业中过度的喧哗已经超出对设计本身的关注。自媒体的兴起起到了很好的信息传播作用，可是这种宣传很容易模糊标准，丢失自身的行业立场。设计师字面解读就是专注于设计的人，就像文学家以文字说话一样，设计师以图纸表达。于此，我和团队的同仁们唯有屏蔽有害，坚定设计认知，磨练专业技艺，时刻警醒自身。

除了忙于设计，您平时有哪些爱好？

如果说一个人自己的爱好就是自己的工作是一件幸福的事情，也许我就是那个幸福的人。除了设计以外，我着力于加深自己的品鉴厚度，将心中山水于闲暇时通过爱好表现出来。习练书法，关注书画，学习古琴，这些都是我的爱好，与设计也息息相关。就像之前提到，我一直在找寻传统的对话途径、心中山水的表达方式。这些爱好恰可辅助我加深对文化对设计的理解。

静谧与
光明的交响
—
路易斯·康
建筑之旅

对于那些低能的建筑师来说，建筑不过是挣钱的来源。而不像它所应该的那样—创造美感和艺术。对我来说，建筑不是事务，而是我的宗教，我的信仰，我为人类幸福、享乐而为之献身的事业。

路易斯·康

3 岁烧伤面颊，50 多岁职业生涯才得以起步，73 岁出差途中遭遇心脏病突发，孤独离世时仍身负重债。这样一位他人看来笼罩着悲剧色彩的建筑师，却称自己的建筑生涯为喜悦的一生。二十世纪的建筑浪潮中，他的建筑突破了时代的局限，传承下来的是超越时代的永存风格。

十九世纪末，芝加哥学派建筑师沙利文宣扬"形式随从功能"的口号，认为建筑设计应由内而外、须反映建筑形式与使用功能的一致性。随之而来的是现代功能主义建筑浪潮。但激进的功能主义主要针对建筑使用最基本的要求，无法解决纪念性的问题。

康认为建筑传递给人们的惊喜，绝不可能随意产生或是因为建筑师有意加入，之所以建成后能让人感受到，是因为它本存在于设计之前。康通过建筑表达着他的认知、他的欲望、他的喜悦、他的悲怆，表达着他和世界、和他人，甚至是和自身的关系。

建筑不可能说谎，身临其境的体验者感受着惊喜的同时也都会有认识了建筑师的感觉，就像亲眼看见了设计过程一样。有些感受很难用语言描述，因理解差异，这趟旅途也一直伴随着与友人的辩论。这篇游记主要由个人主观的入境者角度去剖析康的建筑特征，并试图揣摩他的建筑哲思。

Symphony of Silence and Light

—

Journey to Louis Kahn's Architecture

For inferior architects, architecture is for money, not for aesthetics and arts as it should be for. For me, architecture is not a thing, rather, it's my religion, my faith, and the cause I devote my whole life to, for happiness and pleasure of mankind.

Louis Kahn

For others, Louis Kahn's life seemed to be a tragedy. Face seared at the age of 3, without an established career until his 50s, he passed away alone at 73 due to heart attack during a work trip, still in deep debt then. However, Kahn described his career as a pleasant journey. In the architectural movements of the 20th century, his works transcended the constraints of the times with a distinct style of his own.

At the end of the 19th century, Louis Henry Sullivan of Chicago School advocated "form follows function" and argued that architectural design should consider the interior function as prior to the exterior form and achieve the consistency between form and function. Modern functionalism prevails since then. However, radical functionalism mainly focuses on basic functions of architecture, failing to address the commemorative concerns.

Kahn believes architecture delivers suprises. They neither pop up at random nor are deliberately created by architects. They already exist before the architecture is designed. Kahn conveys his insights, desires, happiness and grievances via architecture, explaining his relationship with the world, with others, and even with himself.

Architecture does not lie. When visiting the architecture, the experiencer will not only feel surprised but also get to know the designer, as if they have witnessed the whole designing procedure. Since some feelings are hard to describe and people's understandings vary, my friends and I have debated all along the way about what we saw. In the travel notes, I analyse the features of Kahn's works and try to explore his design philosophy from my perspective.

耶鲁大学美术馆
—
古典与现代的
完美结合

位于美国康涅狄格州纽黑文市的约克街，始建于 1832 年，后建于 1926 年的耶鲁大学美术馆，由当时的著名建筑师 Egerton Swartwout 设计，外观更像一座修道院，散发着中世纪的醇厚气质。

1951 年在耶鲁大学做建筑学访问学者和建筑评论人的美国建筑师路易斯·康接手美术馆新馆的设计，新馆与原有哥特式古典建筑的老馆相连，外立面简洁、古拙，大面积砖墙与历史悠久的古典建筑形态寻找到视觉重量上的平衡，同时被钢制构件精心分割的玻璃面亦标明其自身的时代身份。在朝向主要街道的立面上没有开窗，大理石的水平分割线延续了老美术馆的哥特建筑大窗比例，同时暗示了内部楼层分割和空间构成，显得克制而朴素。一个放大型的"迷宫格式"入口，从连接处侧方向步入，拾级而上经历灰空间抵达美术馆大厅，体验感层层递进，态度谦卑，使用高效。

内部空间由垂直交通和对称式展厅组成，与原建筑的连接空间配合梯段式下沉庭院组合，这正是康后来提出的"服务"与"被服务"空间的基本形态雏形。垂直交通（服务空间）被布置在中轴线位置，两侧衔接展览空间（被服务空间）。平面布局呈三角形的步行楼梯被放置在圆柱形清水混凝土空间内，由屋面延伸至地下，并在底层形成架空，黑色石料制成的踏板与支撑结构形成咬合，一个看似沉重的垂直交通变得轻盈。而为此空间注入灵魂的另一个端点，正是以三角形为母题的梁板将圆形屋面托举而上，形成立面断离，又似是被环形光线分解的几何体在黑暗中生长出该有的姿态。不同于对光线的粗野支配，这里的建筑与光共同作用形成了一个赋予精神的场域，光成了诱因，为了纪念性！

被看作是被服务空间的展厅部分，楼板采用三角形带肋混凝土结构，将照明、给水等机电系统全部隐藏其中，同时也避免了在建筑之上装饰天花，功能与艺术精妙结合。塑造出大跨度空间的同时尽可能削弱管线对建筑空间的负面作用，将其全部收纳，并形成美妙韵律，与服务空间顶部三角形母题呼应。照明灯光安置在三角梁盖内部，灯光经过折射雾化更显柔和。

柯布西耶在处理一系列建筑时，为了加强空间的流动会倾向于采用圆形柱，这种手法在康的建筑中几乎无处寻觅。与"墙前柱后"的柯布及"柱前墙后"的密斯不同，康不会将柱与墙区分对待，他认为"一排柱子就是一排不完整的墙，而不是别的什么"。如此，在他塑造的空间里配合顶部大跨度梁网的是"断裂的墙体"——方柱，三角梁网与方柱间结合建构出稳定的结构空间关系。

在对内部细节的打磨中，楼梯台阶大理石的处理、混凝土墙面的分缝、栏杆的精心设计都显示出建筑师对于建筑细部的热情。康用现代的建筑理念结合古典哲学、诗化的建筑语言做出了完美的建筑诠释，并为自己树立了第一座丰碑。

Yale University Art Gallery
—

Perfect Fusion of
the Classic and Modern

The Yale University Art Gallery, founded in 1832, is located at York Street, New Haven, Connecticut, US. The old building, designed by famous architect Egerton Swartwout and established in 1926, looks more like a monastery in Medieval Times.

In 1951, Louis Kahn, then a visiting architecture scholar at Yale and an architecture critic, took the job to design the new art gallery, which an extension next to the old Gothic one. The Kahn's building, with a simple and primitive facade and large-area brick walls, matches the time-honored classic old building in visual weight. Meanwhile, the glass curtain walls, exquisitely segmented by steels, are a label of the modern age. The facade facing the main street is windowless. Marble dividing lines cut the glass facade into several big pieces. The design follows the big-window style typical of Gothic buildings and indicates the interior space composition. All show restraint and simplicity. The entrance is designed like an amplified maze. We entered the building from the side of the junction, climbed stairs, walked through the grey space and arrived at the gallery exhibition hall. Everything was unfolded step by step in a humble and efficient way.

The interior space contains vertical traffic facilities and symmetric exhibition halls, jointed by the connecting space and sunken garden, forming a prototype of the "served and servant spaces" proposed by Kahn later. The vertical traffic facilities (servant space) are on the vertical axis, connecting to the exhibition halls (served space) on both sides. The cylindrical structure made by fair-faced concrete is the main staircase with a triangular stairwell inside, extending from the roof to the ground with bottom elevated. Stair steps made by black stones

seamlessly connect with supporting elements. Thus a heavy vertical traffic structure becomes light. Beams and slabs also in triangular shape lift the roof, forming a separated facade, which also seems to be a natural result of different light distributions. The structures, working with light instead of governing it, create a harmonious ambience. Light becomes the trigger, for the memorial!

In the exhibition halls, the floor slabs of the served space take the form of triangular concrete rib structure and house lighting system as well as water supply system, sparing the trouble to design additional decorative ceiling. A fusion of function and art! A long-span spatial structure is created and the negative effects of pipes and wires are reduced as much as possible by hiding them in the structure. Meanwhile the whole design echos the triangle theme. Lighting facilities are installed inside the triangular roof beams, making the reflected lights become softer.

Cylinders were always adopted in Le Corbusier's works to create flowing space while not to be seen in Kahn's design. Different from Le Corbusier and Ludwig Mies van der Rohe, Kahn treated columns and walls equally and believed that a line of columns was a line of incomplete wall, not anything else. Thus in his space, square columns, the "broken wall", match the long-span spatial structure at the top. The triangular beams and square columns together develop a stable structure.

All details in the interior space, such as marble stair steps, seams on concrete walls, exquisite handrails, demonstrate the architect's zeal for details. Having integrated modern values with classic philosophies and poetic architectural language, Kahn offered his understanding of architecture as well as established a monument for himself.

耶鲁大学英国艺术中心
—
不朽的遗世之作

与耶鲁大学美术馆临街而立，英国艺术中心完成于 1974 年（康逝世后），1977 年 4 月 19 日向公众开放。作为康人生的最后一个作品，其内部空间处处透露着对于建筑"透明性"的终生探索。

隔街相望，建筑外观传递出内省的精神性特征。不同于其他作品中典型的混凝土和砖作为主要构筑物料，艺术中心采用磨砂钢板和反光玻璃以模块化的组合方式，填充于被立面柱网均匀分割的方块中。主入口被放置在建筑一角的模数退让空间内，和他本人一样，外表谦卑，建筑内部却精美巧妙。光亮的橡木墙板和石灰华地面通过玻璃正面的过滤，光线在内庭的墙壁与地面留下晕染的辉光。

与成名作耶鲁大学美术馆相同的是，步行交通依旧被放置在圆柱形空间内，顶面采光由半透明玻璃砖排列出的四边形阵列，对由顶面引入的自然光进行二次分配，更为深入地解读了光的可能性。展览区

顶面大尺度的预制空腹梁结构将管线和设备收纳其中，预留出完整的采光顶棚，让人不禁惊叹于这位建筑诗哲的美学认知及技术水平。康着重细部，但精美的细部并不会产生喧宾夺主的感觉，细部之间相互组合而成，建筑整体与光线配合营造出统一的场域氛围，使人沉醉其中直至遗忘建筑本身。

Transparency —— 透明性，界面之间相互依存的关系，在艺术及建筑领域都占有极其重要的位置，一度成为评价艺术作品的标准。康在这个项目中通过对秩序层次的梳理以及立面叠加营造的透视现象，将空间背后的空间及更深处传递到我们面前。垂直方向仍然沿袭了虚实排列及几何叠加的处理手法，维度之间的相互依存使我们在这样的透视关系中不再只局限于面对面的关系。这里，人与空间相互作用，空间与潜空间相互作用，启发出无尽遐想。

纪录片《我的建筑师·寻父之旅》中，有人这样评价他"也许他就是被做成矮个子、丑陋、声音沙哑的犹太人，而且不善于与人相处……这样才让他能够探索内在。"也许他就是被做成这样，才会执着于对美的内在追求。

Yale Center for British Art
—
Monumental Masterpiece

Right across the street is the Yale Center for British Art, which was finished in 1974 after Kahn's death and opened to the public on April 19, 1977. As his final work, every inch in interior space tells Kahn's lifelong exploration of the "transparency" of architecture.

Looking at it from the outside, I feel a sense of self-reflection. Different from other typical Kahn's buildings with concrete and bricks as main materials, it adopts the combination of frosting steel plates and reflective glass to be filled into the squares on the facade separated by posts. The main entrance retreats to a corner, as modest as Louis Kahn. However, the interior space is another world, elegant and exquisite. Sunshine penetrates the glass and gently touches the polishing oak wallboards and travertine floors.

Similar to the Yale University Art Gallery, there's also a cylindrical staircase. Semitransparent glass bricks are arranged into a quadrangle on the ceiling, allowing the sunlight to penetrate. The large precasting open web girder houses pipes, wires and facilities, leaving adequate space for natural lighting. I am amazed at Kahn's aesthetics and skills. He is fastidious. The exquisite details coordinate rather than compete with each other. Together they create a harmonious ambience which fascinates the people so much that they forget themselves and the architecture.

Transparency is about the interdependence between dimensions of structures. In the field of art and architecture, it occupies an extremely important position and once was the standard for evaluating art works. In this project, through combed order of dimensions and the perspective phenomenon created by superimposed elevations, the deeper space behind the space come before us from afar. The vertical direction adopts the combination of virtuality and reality, geometric superposition, and interdependent dimensions, extending the communication beyond the face-to-face interaction . Here, visitors interact with space, and space interacts with latent space. Inspiring!

In the documentary *My Architect: A Son's Journey*, an interviewee said that "Maybe he (Kahn) was made into a short, ugly and hoarse-sounding Jew, and he was not good at getting along with others… it made him go internal". Maybe that's why Kahn was so committed to pursuing aesthetics.

理查德
医学研究中心
———
竖向秩序

费城，一个看似凋零的城市，而在当年的建筑界，因为"理查德医学研究中心"的落成催生了由康领衔的"费城学派"的诞生，并引发了"功能"和"形式"的长久讨论。实验楼于1961年落成，两年以后，又在医学楼旁建起生物楼。这是康在宾大学习执教过程中唯一建造的建筑。建筑落成后因为眩光的实验室窗户被使用者诟病的同时，却受到了建筑师们的极大推崇，并被沙利文称赞为"结构的胜利"。

宾夕法尼亚大学内的建筑似乎很容易让人联想成实验室，寻找"理查德医学研究中心"颇费周折，几经反转，在一片绿荫的簇拥下塔楼露出挺拔的一角，依旧是纪念性的气质。

平面布局似有一种隐形的正交网格柱体布局藏匿其中，使实验楼划分为九宫格形式，旨在复活被现代构成解体的平面中心，体现"拉丁十字的高度对称性"，主轴节奏明晰。单元入口在正交网格的控制下，产生了井字形的平面承重布置，从而解放了角部，用塔楼加强被服务空间的轴线，辅助空间被单独分离出来，沿工作室的轴线布置。北端额外增加一座塔楼，用以强化北向工作楼作为入口部分的标识性，体现了北端作为入口的独特地位。

康巧妙地把工作室、实验室、动物研究、管理、办公等内容分布在一种以"社区单元"为概念的竖向序列空间中，高耸的塔楼串联各"社区单元"形成整个建筑，不同空间的独立意味着结构上的解体，但仍然能感受到动态中的对称性，平面的轴线关系非常清晰。竖向排布的塔楼强调立面竖轴形态，传递给人们强烈的视觉体验。T型窗将窗的采光与通风需求分别解决，对于立面的完整性起到了积极的作用，形成了建筑师鲜明的个人风格。康曾经说"我不喜欢线路，我不喜欢管道，我是完完全全恨透了他们，但是正因为我恨透了它们，我觉得它们应该被给予特定的位置，如果因为厌恶弃之不理，它们就会反过来侵袭建筑"。这栋建筑中水平的空腹梁与竖向塔楼的配合为解决管线问题提供了完美的方式。康将他"恨"的管线安置进了"爱"的塔楼中。

Richards Medical Research Laboratories
—
Vertical Order

Philadelphia seems to be bleak. However, it is where the Philadelphia School, with Kahn as a leading representative, was born, thanks to the Richards Medical Research Laboratories which triggered the lasting discussion about "function" and "form". The laboratory building was finished in 1961. Two years later, the biological research laboratory was built beside it. The Richards Medical Research Laboratories were Kahn's only project during his tenure at the University of Pennsylvania, criticized by the occupants due to the dazzling windows but acclaimed by architects such as Sullivan who eulogized it as the victory of structure.

Buildings at the University of Pennsylvania all look like laboratories. It took us quite a long time to find the Richards Medical Research Laboratories. Towers thrust from trees, and the design of which conveys a memorial touch.

There seems to be an invisible orthographic grid structure in the layout, which divides the laboratories into nine grids. It intends to present the perfectly symmetrical Latin Cross and highlight the main axis. The unit entrance, confined by the orthogonal grid, has a #-shaped load-bearing layout, which frees the corners. Towers reinforce the axis of the served space and separate the servant space which is distributed along the axis of the studios. The northern end has one more tower, a symbol revealing its uniqueness as the entrance.

Kahn arranged the studio, laboratory, administration office, etc. in vertical space with the conception of "community". High towers are connected to all "community units" as a whole. Though different communities are quite independent from each other, there's still a sense of dynamic symmetry with distinct two-dimensional axes. Towers with clear vertical lines highlight vertical axes in an impressive way. T-shaped windows meet the requirement of lighting and ventilating, and reinforce the wholeness of the architecture as well as present distinct personal style. Kahn said that he didn't like wires and pipes, he hated them, but it was because he hated them that he thought he should arrange special places for them, otherwise they would erode the architecture. The horizontal open web girders and vertical towers perfectly hide those "annoying" wires and pipes. Kahn arranged the wires and pipes he "hated" in the towers he "loved".

索尔克生物研究所

—

混凝土与砾石的神庙

一位年轻的建筑师引领我们开启了这一段让人着迷的旅程，非常凑巧的是，他从业的事务所正是康的弟子所创办的。当我们驱车四个小时穿越沙漠到达目的地时，这座久负盛名的建筑正在进行维护工作。不知是不是被我们不远万里前来的诚意感动，管理人员竟然破例允许了我们的参观。

索尔克生物研究所位于美国加州圣迭戈市拉霍亚（La Jolla，California）北郊悬崖上，西向大海，面临壮观的大西洋。基地内是高台地和峡谷，环境深幽，和其他世界缺少联系。据说在这种孤寂的环境下，当年索尔克先生为了能将科学家留在这里，对康的设计要求是"我希望能建造一栋最美的建筑，即便是毕加索来了也不愿意离去"。而康确实做到了这点。在我们以往的认知中，建筑艺术归属于一种独特的艺术范畴。建筑作品由于特殊的体量关系、复杂程度、工艺技术以及材料特性的制约，用艺术品的表达几乎不可能实现。索尔克研究所突破了建筑艺术性的界限，传达出强烈的精神力及建筑信仰，利用美的创造来实现与现实世界的抗争。

由中轴侧边大门进入，经过一组对称式建筑后，沿整块石料切割而成的踏步向上，进入一个类似于"桥"的领地，研究所最重要的建筑群已尽收眼底。其形态更像是意大利阿西西的修道院，建筑主体全部采用钢筋混凝土结构，木材作为围护部分与混凝土间采取剥离的手法隔断处理。混凝土由于混合当地大理石的粉末而呈现出轻微的粉红色调，偏暖的色调带走了混凝土的冰冷质感，与木材相互和谐，与基地色调呼应，仿佛由场地生长出一般。两组研究楼沿中庭广场相互对称，侧向大海，仿佛海边神庙宁静致远，超越物质的空间精神性映射出观者内心，唤醒共鸣感受。

中心广场上正直的水线居于轴线位置，强调了空间的对称和指向性。水流始于矩形的源头并指向尽端的太平洋，将可量度的距离引入不可量度的想象，源头和归处的含蓄隐喻暗示生命的起始。广场两侧的建筑柱廊限定出中央的大厅空间，区分出空间的主次，同时形成空间纵向的方向性，海的声音，风的声音，鸟翅拍打的声音，脚步的声音，陪伴着沉默的建筑。广场阶梯式向海面下沉，场地的高低改变着我们的视点，上下间形成不同的水平变化，邀请人们接近，接近，更接近海面，最终海天一色。

在最初的设计中，庭院被一个狭窄的水渠平分，沿着水渠种植了两排意大利柏树，直到康邀请路易斯·巴拉干为当时还未完成的中庭广场提些建议。康经常提及这样一段经历："当他进入这个空间的时候，他走到混凝土墙的旁边，抚摸着它们，并表达了对它们的喜爱。当他的视线穿过这个空间望向大海的时候，他说，我不会在这里种一棵树或一片草。这里应当是一个石头的广场而不应是个花园。我和索尔克四目相对，都觉得这个想法非常正确。他感觉到我的赞赏，又高兴地补充说，如果你把它建成一个广场，将会获得一个立面——一个朝向天空的立面。"

当年，建成后美轮美奂的索尔克生物研究所，不但留住了世界上最顶尖的科学家，也留下了作为康的助手参与整个设计的杰克卡斯特蒙，并且杰克卡斯特蒙在有生之年都不打算离开。

建筑不可能说谎，身临其境的体验者感受着惊喜的同时也都会有认识了建筑师的感觉，就像亲眼看见了设计过程一样。旅行虽然短暂，却使人感受良多，而康的建筑信仰亦将继续引领我们前行，尽管路程艰辛，吾亦甘之若饴。

Salk Institute for Biological Studies
—
Temple of concrete and Gravel

A young architect led us through the fascinating journey, who happens to work for the company founded by one of Kahn's students. After a four-hour drive through the desert, we were told the architecture was under maintenance. Maybe moved by our sincerity of coming from afar just for visiting the famous buildings, the administrative staff let us in.

Salk Institute for Biological Studies is located on the cliff in La Jolla, California, facing the magnificent Pacific Ocean. Surrounded by plateaus and valleys, the Salk Institute is isolated from the rest of the world. How to keep scientists in such a lonely place? One of the requirements of Dr Salk is that the architecture should be "worthy of a visit by Picasso". And Kang really did it. Architectural art is thought to be unique. Due to the huge size, complexity, procedures and technology, and material properties, it's very different from handicrafts. The Salk Institute is freed from the limits of architectural art and conveys strong commitment and faith, using aesthetics to resist the lures of the real world.

Through the gate beside the axis, past symmetrical buildings, along the upward stairsteps, we arrived at a bridge-like place and saw the main complex. It was inspired by the San Francesco d'Assisi Monastery. The main body consists of reinforced concrete structures with wood fences, and the wood as the enclosure part is separated from the concrete by peeling. The concrete has a light pink color because it is mixed with local marble powder. Warm color takes away the aloofness of concrete, echoing the color of wood and the surroundings, making the whole architecture look as if fresh from the earth. The two symmetrical groups of buildings along the central plaza in parallel with the ocean are like the pantheon at the seaside, tranquil and elegant. The spatial spirituality beyond materials evokes resonance and reaches the inner world of the visitors.

The straight water channel locates at the axis of the plaza, emphasizing the sense of symmetry and direction. The flow starts from a rectangular source and goes into the Pacific Ocean. The unmeasurable imagination space is introduced into measurable distance. Where does life start? Where does life end? The design is a metaphor. Pillars of the buildings on both sides of the plaza define its space, distinguish the primary and

secondary of space, and also point out the longitudinal direction. The silent architecture is accompanied by the sound of the ocean, wind, wings of birds, and footsteps. The sunken plaza got us closer and closer to the seaside, step by step. The ocean was inviting us. Finally, the sea and the sky merged into one.

For the initial design, the square was evenly divided by a minimalist water channel, along which were two rows of Italian cypresses. Luis Ramiro Barragán Morfín was invited to give some advice about the unfinished central plaza. Kahn often mentioned the experience. "When he (Barragán) entered the space, he went to the concrete walls and touched them and expressed his love for them, and then he said as he looked across the space and towards the sea that he would not put a tree or blade of grass in this space. This should be a plaza of stone, not a garden. I (Kahn) looked at Dr. Salk and he stared at me and we both felt this was deeply right. Feeling our approval, he added joyously and said that if we made this a plaza, we would see a facade-a facade to the sky."

Many world-renowned scientists have settled down here and don't want to leave since the breathtaking architecture was established. Jack Mc.Callister, Kahn's assistant then and involved in the whole design, haven't left here since he moved in and will never leave in his lifetime.

Architecture does not lie. When visiting the architecture, the experiencer will not only feel surprised but also get to know the designer, as if they have witnessed the whole designing procedure. The journey was short but colorful and inspired me a lot. Kahn's architectural faith and philosophies will continue to lead me forward. Despite bitterness and difficulties along the journey, I enjoyed it.

亚平宁
建筑之旅 13 日
—
关于
卡罗·斯卡帕

卡罗·斯卡帕
Carlo Scarpa

莫霍利·纳吉认为，"人们永远无法通过描述来体验艺术，解释与分析不过是一种知识的储备，不过这能鼓动人们与艺术进行亲密接触"。卡罗·斯卡帕，一位没有追随者的大师，他的作品必须身临其境才能感受到令人流泪的力量。

一直非常抗拒写游记，别人的游记诚实说也很少去读。旅行是非常寻常、私密的行为，就像不会刻意记录每天的食谱，途中经历也应该随着时间的消化，留下的被吸收，默默发挥着影响，其余则被磨灭、忘却、遗失在角落。经常出去走走，也参观过不少优秀的建筑作品，至今只有两次面对建筑瞬时泪流满面。一次感动于黄昏光晕中的朗香，另一次便是被 Brion 墓园强大的精神力量撼动，流连忘返、迟迟不肯离去。亚平宁这十三日起初是为米兰世博会和威尼斯双年展安排的时间，却被这个对细节追求到类似有强迫症状的老爷子震撼了心灵。末了还是觉得应该写点什么以纪念这位大师送给我的惊艳。

拜伦曾经这样赞叹："忘不了威尼斯曾有的风采，欢愉最盛的乐土，人们最畅的酣饮，意大利至尊的化装舞会。"这个戴着精美面具的城市为绘画和艺术创作提供了阳光灿烂的充满自信的环境。由于其特殊的地理位置，从古至今威尼斯一直保持着其商业的繁荣、宗教的自由、文化的合璧。百岛城古迹众多，到处是画家、作家、音乐家和建筑家留下的著作。20 世纪，南欧地区受印象主义及后印象主义风格影响较大，使得艺术家们对作品的表达更关注艺术本身，偏重艺术形象的感性表现。卡罗·斯卡帕在这座水城度过了一生中的绝大部分时间，他的作品扎根在文化背景下，沁染着传统气息。风格派影响了他的构成系统，东方文化的影响使得斯卡帕的建筑更具有区别于同时代其他建筑的沉静雅致。

A 13-day Architectural Tour in Apennine

Tour in Apennine

—

About Carlo Scarpa

Moholy Nagy believes that "One can never experience art through description. Explanations and analysis can serve at best as intellectual preparation. But they can encourage people to have close contact with art." The power of the works of Carlo Scarpa, a master without followers, can not be felt without being personally on the scene.

I have always been resistant to writing travel notes. Honestly, I seldom read other people's travel notes. Travel is a common and intimate act. Like we never record our daily recipes, there is no need to record our journey. The experience on the way should be digested with time; the left behind will be absorbed, playing a role in our life silently;

and the rest will be obliterated, forgotten, and lost in the corner. I have travelled a lot, and have visited many excellent architectural works. Up to now, only two of them have moved me to tears. One is the Longchamp in the halo at dusk, and the other is the Brion Cemetery, whose strong spiritual power struck me so hard that I forgot to leave. The 13 days in Apennine were initially arranged for the Milan Expo and the Venice Biennale, but I was impressed by Carlo Scarpa, who pursued details like an OCD patient. In the end, I felt that I should write something to commemorate the amazing feeling this master brought to me.

Byron once wrote, "Nor yet forget how Venice once was dear. The pleasant place of all festivity. The revel of the earth, the masque of Italy"! This beautiful city with an exquisite mask provides a confident environment for painting and other artistic creation. Because of its special geographical location, Venice has maintained its commercial prosperity, religious freedom and cultural harmony since ancient times. There are many historic sites in Baidao City, where works left by painters, writers, musicians and architects can be found everywhere. In the 20th century, Southern Europe was greatly influenced by impressionism and post-impressionism, which led artists to pay more attention to art itself and to the perceptual expression of artistic images. Carlo Scarpa spent most of his life in this city of water. His works are rooted in a cultural background with traditional flavors. De Stijl influenced his composition system, and the influence of oriental culture made his works more elegant than other contemporary buildings.

Castelvecchio
城堡
博物馆
——
半遮面的遗憾

Castelvecchio 位于维罗纳阿迪吉河畔，始建于中世纪。14 世纪由统治者斯卡拉家族建造完成；18 世纪改为威尼斯陆军学院；19 世纪遭遇战火摧残，在居住院落建造了一座 L 形的新古典建筑用作防御工事及营房；一战结束后，于 20 世纪 20 年代改为陈列该地区中世纪艺术的博物馆；20 世纪 50 年代中期，卡罗·斯卡帕着手主持该项目的设计，终于于 20 世纪 70 年代完成了整个工程。

由于周一博物馆闭馆维护，我们只能从外部领略他的风采。显然，斯卡帕在改造中，力求通过梳理建筑的片段，使各个历史的层面真实地展现。传统的威尼斯磨光硬水泥抹面、熟石灰抹面，与原有的粗糙墙体带着自身的历史特征形成差异效果，毫不掩饰地在同一时空内互相凝视。在处理交界关系上，斯卡帕使用了惯用的驳接手法，留设出一道低于建筑表面的缝隙，突显临界，强调新旧。在桥、建筑交接处的小庭院，斯卡帕将 Cangrande 骑马雕塑作为空间焦点，放置于庭院上空的水泥平台之上，掀起一个局部空间的小高潮。

斯卡帕保留了威尼斯人的特有习惯，在同一幢建筑上，即使某一个部位的外皮曾经脱落了，或者窗户的位置发生了变动，他们也不会掩饰过往残缺，反而会清晰地留下历史上曾经经过修缮的痕迹，于是形成现在墙面的斑驳质感。天空湛蓝、日光强烈，质感粗糙的历史墙面衬托出玻璃宝石般的光泽，赞叹之余，谁又知晓这位闻名于世的建筑艺术大师早年也曾经是威尼斯最好的玻璃匠人？

威尼斯本土盛产一种微微泛着粉红的大理石（搜集资料猜测为 Istrian stone，尚不确定），自古当地的大部分石刻艺术都使用这种材料作为创作的物理载体。斯卡帕通过现代感极强的黑色钢质的框架，将粗糙的大理石牌碑排列展示。色彩的反差、材料差异性的并置形成了神奇的视觉效果。通过突出强烈的年代差异，塑造出建筑空间氛围的历史层次。

Castelvecchio Museum
— Half-concealed Beauty with Imperfection

Located on the banks of the Adige River in Verona, Castelvecchio was built in the Middle Ages, completed by the Scala family, the then ruler, in the 14th century, and transformed into the Venice Army College in the 18th century. In the 19th century, after it was devastated by war, an L-shaped neo-classical building was built at the courtyard to serve as a fortification and barrack. After the First World War, it was transformed into a museum in the 1920s to display medieval works of art in the region. In the mid-1950s, Carlo Scarpa started to take charge of its design, and in the 1970s, the whole project was finally completed.

Because the museum was closed for maintenance on Monday, we could only appreciate it from the outside. Obviously, during the transformation, Scarpa strived to present all historical periods through every fragment of the building. The traditional Venetian polished hard cement plaster and lime plaster stood in contrast with the original rough wall with historical characteristics, and both of them staring at each other in the same space at the same time. In dealing with the borderline, Scarpa's traditional method of connection was adopted and a gap was left below the surface of the building to highlight the dividing line between the old and the new. In the courtyard at the junction of the bridge and the building, Scarpa placed the Statue of Cangrande on the cement platform as a focus, setting off a climax in this part of the space.

Scarpa kept the unique habits of the Venetian, who would not conceal the imperfection of the past even if the plaster had peeled from the wall, or the position of windows had been changed, but would keep the traces of repairs. The mottled wall nowadays is thus formed. Under the blue sky and bright sunshine, the historic rough wall is the perfect foil for the jewel-like luster of the glass. When people admire, who knows that this world-famous architect was also once the best glassmaker in Venice?

Venice is home to a slightly pinkish marble (Istrian stone by guessing based on gathered material, but uncertain), which is used as the materials for most of the local stone carvings since ancient times. Scarpa showed the rough marble tablets in sequence in the very modern black steel frames. The contrast of color and the juxtaposition of different material creates magical visual effects. By highlighting the dramatic differences in ages, the historical levels of the space are shaped.

Brion 墓园

—

生与死的终极浪漫

Brion 墓园位于特莱维索附近的桑·维多。基地面积约为 2 200 平方米，大体呈 L 形布局。与桑·维多公墓结合成统一区域。在墓园公路入口下车，沿着碎石林荫道行走约 15 分钟，墓园的矮墙渐行渐近。矮墙略向内倾的形态，暗示了墓园内部主题与现实世界的距离。两段倾斜围墙交接的部分选用了类似中国"喜"字装饰性镂空图案进行虚体转折。

去往墓园参观的人，入园之后，徜徉之前需要稍停顿，清空已有的对建筑的认知，方能享受斯卡帕的创造力之美。笔直的道路、几何化的布局、人为规划的水系将整个园区规划成水池、墓地、礼拜堂三个部分。在这个看似由西方数学逻辑思维主导营造的空间里，意外的融合了东方园林的特点，使用了絮语化的笔触，以漫游似的布局将情怀缓缓叙述。感受不到对死亡的暗叹和恐惧，也体会不出对生命的向往和渴望，只剩下巨大的安宁。在这里，生死间不再有巨大的鸿沟，死亡并没有被设定为人生之外的东西，在这里，死作为生的一部分得以永存。

在斯卡帕的设计生涯中，赖特和路易斯·康两位大师对他产生过很深远的影响。墓园这部作品更贴近康的气质。康一直拥有深沉的古典气质，斯卡帕亦是古典的，他的思想早已超越了功能主义的设计，用画笔缓缓描述着自己仿佛来源于中世纪的极度浪漫。他将墓主人夫妻的棺木以桥覆盖，确切来说

可能更贴近于一种混凝土拱券。拱券下的棺木微成角度对放，"两个人在生前相互敬爱，死后也应该在地下相互致敬"，他这么以为。

在人们惯性思维中要默认没有路径的园林是很一件困难的事情。墓园中除了主要引导路线，几乎没有设置任何小径，这于现代的思维而言是完全不可思议。在这里可以体验到一种类似于日本枯山水的氛围，那是一个不需要打扰的空间，人们从远距离观赏便可，各种"死径"的设定配合景框元素的设计，使人的视线到达身体所不能及的地方。

斯卡帕的建筑与现代主义建筑拥有很多共同点。他们都采用纯粹的几何语言，都会采用简洁的方式塑造建筑，建立建筑的形式秩序。而斯卡帕自身最具特点的，是他丰富的图像世界，图像在形式上的自主性和非指称性是其设计的重要特征。斯卡帕的形式充满着符号的意义，并没有特意的具体隐喻或所指，就像他镂空的"喜"字并不带有其实际中文意义，图形只是自我表达的主体，是建筑形式的表达需要。

在这里我们能欣赏到贯穿于斯卡帕大部分作品的形式母题之一——5.5厘米乘5.5厘米的模数线脚。檐口、腰线、柱体、水池、花坛等等部位，斯卡帕反复使用这种叠涩线条作为结构、节点的装饰。在室内空间，斯卡帕依然不断变化着材料，改变尺寸、位置和尺度，对此母题反复咏唱，以达到整个体系的统一性和完整性。

此外便是著名的"相交的圆环"。双圆叠加，一蓝一红，一边海水一边火焰。同行好友理解为生生不息的阴阳之说，私略觉勉强。只觉得确实感知到有这样两只眼睛，暗示着另一个空间的存在，引导我们从生的现实洞悉到河流的另一端。

对于墓园里最重要的建筑—教堂，斯卡帕花费了很多笔墨。进入教堂要经历一个非常迂回的过程，经历一系列材料的过渡、空间的转折，最后通过一个上不封闭的圆形洞口，整个设计有预谋地给心理意识设置了层层虚幻的关卡，仍然吟唱叠涩主题。墙体的剖面化体现出墙体的厚度感，混凝土的石板化质感处理勾显出建筑表面的质感。阳光穿透深厚的墙体落在地面，漫射在空气中，安静、清冷、富有神性。

Brion Cemetery
— The Ultimate Romance of Life and Death

Located in San Vito near Treviso, the Brion Cemetery covers an area of about 2 200 square meters in an L-shaped layout and forms a unified area with the San Vito Cemetery. Get off at the entrance to the cemetery and walk along the gravel road for about 15 minutes, you will find the low wall of the cemetery gradually approaching. Its slightly-inward slope implies the distance between the internal theme of the graveyard and the real world. A decorative hollow pattern, which is similar to the Chinese character " 喜 ", is used as a transition at the junction of two sections of the inclined wall.

Those who visit the cemetery need to pause a little before wandering to empty their existing knowledge about architecture, so as to enjoy the beauty of Scarpa's creativity. The straight roads, geometric layout and artificially planned water system separate the cemetery into three parts: pool, cemetery and chapel. The unexpected blending of the features of Eastern gardens in this space, which seems to be dominated by Western mathematical logic, expresses the thoughts and feelings of the designer in a whispering way. One cannot feel the lament and fear for death, or the yearning and longing for life, but only peace and tranquility. Here, the huge gap between life and death no longer exists. Death is not set as something incompatible with life, but can remain forever as part of life.

Wright and Louis Kahn had a profound influence on Scarpa in his career as a designer, and the style of the cemetery is more similar to that of Kahn. Kahn is a designer of classicism, so is Scarpa, whose ideas had long surpassed the designs of functionalism. Scarpa used painting brushes to slowly describe his ultimate romance, which seems to be derived from the Middle Ages. He covered the coffins of the tomb owners, a couple, with a bridge, specifically speaking, a concrete arch, and put the coffins face to face with a slight angle, as he believed that "Two people who love each other when they are alive should greet each other under the ground after they die".

It is quite difficult for people to accept a garden without a path in their inertial thinking. Except for the

main guiding route, there are almost no paths in the cemetery, which is totally unimaginable in modern thinking. Here you can experience an atmosphere similar to the Japanese rock garden, which is a space that needs not to approach, but only needs to view it from a distance. Various "dead paths" combined with frame elements enable people's sight to reach places where the body cannot.

Scarpa's works share a lot in common with modernist architecture. They all use purely geometrical languages and simple methods to shape buildings and establish form and order. The most distinguishing characteristic of Scarpa is his richness in image. The autonomy and non-referentiality of an image in its form is an important feature of his design. The forms of his design are full of symbolic meaning, but have no specific metaphor or signified, just as the hollow " 喜 " does not contain its actual Chinese meaning. Images are only the subject of self expression and the expression of architectural forms.

 Here we can admire one of the form motifs that run through most of Scarpa's works—moldings with a modulus of 5.5 cm×5.5 cm. Such corbel moldings are repeatedly used as a decoration of structure and nodes on cornices, waistlines, columns, pools, flower beds, etc. In the interior space, Scarpa kept focusing on the motif and constantly changed the material, size, position and scale to achieve the unity and integrity of the entire system.

We can also admire the famous "intersecting circles". The two circles overlap each other, one blue and one red, symbolizing seawater and flame respectively. My friend who visited with me interpreted them as the everlasting yin and yang. I thought it was a little far-fetched, but I did feel that they were like two eyes suggesting the existence of another space, and leading us to cross the river between life and death.

For the most important building in the cemetery, the chapel, Scarpa spent a lot of efforts. The entire design sets up various illusory checkpoints for people's mind, with a series of transitions in material and space, as well as an unclosed circular hole at the end on the way into the chapel. Still focused on the motif, the section of the wall reflects its thickness, and the slate texture of the concrete highlights the texture of the building's surface. With the sunlight shining through the thick wall and diffused in the air, the chapel looks tranquil, chilly and divine.

Stampalia
基金会
—
叠加后的艺术馆

由于时间的原因，与举世闻名的光之美术馆Possagno 失之交臂。Stampalia 基金会的顺利参观多少给受伤的心灵带来一丝安慰。这个于十八世纪建成的宫殿，斯卡帕在两个世纪后完成了对其底层和花园的修复工作。

在园林水系中，迷宫图案这一次成了母题线索。几乎跨越了整个园林的水流形成了一个水的迷宫，水流从源头的方形大理石迷宫逐步跌落，直至尽头的圆形混凝土迷宫。

在建筑处理上，斯卡帕将底层空间整体打造出一层新的表皮，但并不进行完全的封闭，使得人们能够通过外在表皮去观察建筑的原始内里，充分地展现出斯卡帕的驳接艺术。新与旧之间、面与面之间、材料与材料之间都细致地进行了"缝"的处理。这里的"缝"我们尝试从两个方面进行理解。从建筑功法来看，这种技术处理强化了各要素之间的组合感，犹如机器由各个零件组装而成，此时建筑也成为一部巨大的机器。从精神感

受上来看，斯卡帕善于将一些新的材料、新的界面叠加于历史建筑之上，缝隙恰似交叠时间之间的裂痕，成为新旧对话的介质，我们从中窥探，引发强烈的怀旧情绪和历史的责任感。

斯卡帕的设计强调细部，尤其对节点及其装饰性能有着狂热的追求，坚持颂扬"上帝也在细部之中"的论调。他的作品不分室内与室外，不分建筑与家具，全部打上了特有的细节符号。但凡是他的设计作品都具有强烈的可识别性，偶遇的小桥、用于展出艺术品的画架、池边维护的构筑物，都和它的创造者一样固执地展现着自己的个性。

因而，有些人也会产生斯卡帕作为"匠人"比建筑师更贴切的想法。其实早在包豪斯宣言中，便清晰地对艺术家进行了定义："艺术家只是一个得意忘形的技师，在灵感出现并且超出个人意志的那个瞬间片刻，上帝的恩赐，使得他们的作品变成艺术的花朵。"斯卡帕的作品是一种由各种美观共同组合的实体，他用建筑、雕塑和绘画三位一体塑造出美的殿堂。卡罗·斯卡帕枯萎在东方的旅途中，安眠于自己最伟大的著作中，是一位令人敬佩又心生羡慕的先驱。所谓没有追随者，可能只因他无法被追随吧。

Stampalia Foundation
—
A Museum of Superimposition

Due to time limits, I did not visit the world-famous Canova Museum in Possagno. But the visit to Stampalia Foundation brought a little comfort to my wounded heart. The palace was built in the 18th century, and Scarpa completed the restoration of its ground floor and gardens two centuries later.

This time, maze pattern becomes the motif of the water system in the gardens. Water flows across almost the entire garden, forming a labyrinth, and falls from the square-shaped marble maze of the source to the rounded concrete maze at the end.

In terms of the building, Scarpa created a new surface for the ground floor that is not completely closed, making it possible for people to observe the original interior of the building through the external surface, which fully demonstrates Skarpa's art of connection. A "gap" has been carefully left between old and new, surface and surface, material and material. We try to understand the "gap" from two aspects. From the perspective of architectural methods, this technique strengthens the integration among various elements. Just like machines are assembled from various parts, the building has also become a huge machine. From the perspective of mental experience, Scarpa is good at superimposing new materials and

surfaces on historical buildings. A gap is just like a crack between the overlapping time, becoming the medium of old and new dialogues. As we peeped through the gap, strong nostalgia and historical responsibility was aroused.

Scarpa's design emphasizes details, especially the nodes and their decorative function, and insists on praising the argument that "God is in the details". All his works, whether indoors or outdoors, buildings or furniture, are marked with special details, thus are highly identifiable. A bridge that you run into, an easel for displaying artworks and the structures for maintaining the pool all show their individuality as stubbornly as their creator.

Therefore, some people may think that Scarpa is more a "craftsman" than an architect. In fact, it was clearly defined way back in the Bauhaus Manifeto that, "The artist is an exalted craftsman. In rare moments of inspiration, moments beyond the control of his will, the grace of heaven may cause his work to blossom into art". Scarpa's works are a kind of entity composed of various beauties. He used the trinity of architecture, sculpture and painting to create a palace of beauty. Carlo Sparpa, who passed away on his journey to the East and slept the final sleep in his greatest work, is a pioneer admired and envied by people. The reason some people say he has no follower may be because he cannot be followed.

番外

—

威尼斯的商业街道

距离离开威尼斯还有二十分钟的时间，忽然听闻 Olivetti 商店就位于距离当下五分钟路程的不远处，果断弃下行李一溜烟跑了去。在圣马可广场众多商业建筑中，斯卡帕的建筑语言非常独特，易于辨识，可惜似乎已多日没开业经营，只能隔着玻璃领略下其最表象的风采。回国的途中不忘搜集 Olivetti 的资料学习，惊诧地发现虽然经过半个世纪的时间洗涤，这个经典的作品跨越了时间的鸿沟仍保持着它才落成的模样，向世人展现着它优雅的姿态。

不禁想到我们身边的商业街道。因为商业街道是公众活动最活跃也是最脆弱的地方，所以它往往非常容易受到城市变迁的冲击而发生改变。在我们国家，由于种种因素的影响，构成街道的许多历史信息会被冲击、掩盖或者彻底遗失。对一个个街区的推翻重建似乎被我们认为是理所应当的。对比而言，即便是在欧洲以奢侈著名的威尼斯人，对待老建筑的态度也更加尊重、体贴。他们尊重历史的痕迹，并以此为荣。于是，街道的"基因图谱"脉络非常清晰，有时仅仅依靠建筑本身便可以记录人们的改变和环境的变迁。当地人拥有着深沉的历史怀旧情绪，对于已有建筑艺术更充满自信。与之相比，我们轻易将一栋栋建筑推倒粉碎又反复，问题到底出在了哪里？

Another Story

—

Commercial Streets in Venice

When there was twenty minutes left to leave Venice, I suddenly heard that the Olivetti Showroom was only five minutes away, so I left my luggage without thought and ran over. Among the many commercial buildings on the Piazza San Marco, the building designed by Scarpa is very unique and easy to recognize. Unfortunately, it seems that it has not opened for several days, so I could only appreciate the elegance of its appearance through glass. On the way back to China, I collected some information about the Olivetti Showroom to study. I was surprised to find that after the vicissitudes in half a century, this classic work still maintains its original appearance, showing the world its graceful posture.

I could not help thinking of the commercial streets around us. As the most active and vulnerable place for public activities, they are often vulnerable to changes in the city, and will change under the impact. In our country, due to various factors, lots of the historical information that makes up the streets will be impacted, covered or completely lost. The demolition and reconstruction of blocks seems to be taken for granted. In contrast, even the Venetian people who are famous for extravagance in Europe are respectful and considerate to old buildings. They respect the traces of history and take pride in it. As a result, the "genetic maps" of the streets are so clear that sometimes people can record changes of human beings and the environment just by relying on the buildings. The locals have a deep sense of nostalgia and are more confident about the existing architectural art. In contrast, we rashly tore down a building and built another one over and over again. What has gone wrong?

设计感悟之建筑与时装

今天多元化的设计领域里，为什么唯独建筑的美态征服了众多善变的设计师，得以走进时装的灵魂？答案也许是因为建筑设计师和时装设计师们非凡的才华，让他们对美的感受顺利达成共识；也许是服务于共同的观众，肩负同样美化空间、缔造快乐生命的使命，但也可能答案根本不会这么复杂，只是因为美好的建筑和时装，让我们的视觉愉快，心情舒畅，并因此爱上了它们，就是这么简单！

设计中常提到"以人为本"，建筑师不妨把"以人为本"的"人"解释为"身体"，于是便有了身体、服装、建筑的三角关系。遮蔽还是暴露？建筑的矛盾性和复杂性，使之产生了与服装相同的命运。尽管建筑与时装在争夺身体的青睐方面有着竞争关系，两个领域自20世纪80年代之后便越来越多地共享一些相似的策略。

对于服装的柔韧，建筑曾经难以企及，然而通过对刚性材料的再编织，新的组织能够超越材料的固有属性，在更宏观的尺度上获得类似褶皱式织物的柔韧性。90年代后的计算机技术，使得更为复杂的结构和形态成为可能，从而使来自80年代后现代建筑思维的构想从纸上谈兵变为现实。建筑立面成了将建筑与城市，建筑使用者和城市过客分离开来的界面，成了建筑身体的时装，而遮蔽和暴露则成为建筑把握的内容与城市语境之间关系的游戏：有时透明得令内部活动一览无遗，有时像人们展示胸大肌一样展示自己彪悍的结构或设备，有时则以封闭的幕墙或巨幅影像将内外隔离。如果说时装是一门将织物包绕于身体上的学问，建筑和建筑环境需要考虑的则不限于此，它还需要考虑如何组织和重构建筑"身体"本身，即其所容纳的社会组织。建筑更像一门内外兼修的学问，它在立面上的工作则是要设定由内而外的表达方式，而时装则可以从建筑的这种整体把握中取得在内外平衡上的借鉴。

Sentiment of Design: the Architecture and the Clothes

Nowadays, in the increasingly diverse design industry, why is it that only the architecture fascinates numerous capricious fashion designers and makes its way into fashion designing? Perhaps it is because both architects and the fashion designers, endowed with remarkable talents, naturally have a common understanding of beauty. Or it is because they serve the same audience and share the common mission to beautify the space and create happiness. But the answer to this question may not be that complex at all. The elegance of both architecture and clothes makes us feel delightful both at sight and in the mind and makes us fall for them. Maybe the answer is as simple as this!

In designing, the "human-oriented" concept is constantly brought up. If we interpret the word "human" as "human body", then the triangle of human body, clothes and architecture comes into being. To remain covered or to become exposed? The ambivalence and complexity of the architecture make the architect have the same fate as clothes. Although architecture and clothes seem to compete against each other in winning over people's favor, they come to share more and more similar strategies ever since the 1980s.

设计之都的法国和意大利，时装设计师都需要持有建筑师的执业资格，听起来似乎让人费解，然而从 Celine、Balenciaga、MaxMara 填充过的肩型中不难看出设计师们对建筑轮廓的迷恋。关于未来，还有多少种可能性？也许答案就在天才设计师 Zaha Hadid 女士笔下的迪拜歌剧院里，这座梦幻般的建筑有着类似于海浪状的超现实弧线，以及山峰一样的起伏高度，形态自然且凌驾于可见的自然状态之上，并且完美地同周边的环境结合在了一起。人们为未来主义建筑的震撼效果而发出的尖叫声同时会在 Rick Owens 的秀场上响起。相同的不受束缚的设计思维模式让时装与建筑作品有了惊人的相似，锐角度的拼接、反常规的波浪起伏的下摆、高耸的领口，这些具有清晰建筑感的未来主义设计极大地拓展了我们的想象空间。就像在 Hussein Chalayue 的设计作品中，你找不到任何一条墨守成规的破缝线一样。流畅简约的外轮廓，被面料舒展精致的立体剪裁所丰富着，由此带来的视觉上的艺术美感最终成就了时装上的柔软"未来建筑"。

In architecture, it used to be hard to achieve the pliability seen in clothes. But by reweaving the rigid material, the new structure overcomes its intrinsic property and becomes pliable like the pleated fabric at the macroscopic scale. Since the 1990s, the computer technology has made more complex structures and forms possible, turning postmodern architectural thinking of the 1980s into reality. The building facade separates the building from the city, divides the users from the passers-by, becoming the clothes for the architecture body. To remain covered or to become exposed also becomes a game for the designers in balancing the content and the context: sometimes the building facade is so transparent so as to completely expose the inner space; sometimes it exhibits its structure and facilities bluntly just like a brawny man shows up his muscles; and sometimes it isolates itself from the outside world with curtain walls or huge billboards. If we see the fashion design as an art to wrap up the body, the architecture design is more complex: it studies how to organize and reconstruct the body itself—the social organization that the architecture holds. In this way, the architecture design becomes an art that focuses on both the inner and the outer world. Its goal is to find an expression that introduces the former to the latter, and fashion designers can learn from this holist approach to reach a balance of the two worlds.

In Paris and Italy, cities of design, fashion designers all need to hold an architect's certificate, which may sound confusing. But from the shoulder pads that Celine, Balenciaga, MaxMara have designed we can see the obsession that the designers have toward the outlines of the architecture. How many possibilities will the future hold for us? Maybe the answer hides in the Dubai Opera House designed by the design genius Zaha Hadid. With surreal waves and ups and downs like mountain ranges, this building looks natural but at the same time transcends any existing natural forms, fitting in perfectly with the environment. People get thrilled at the sight of this stunning futuristic architecture, and such thrill can also be heared on the runway of Rick Owens. Similar design logic generates a striking similarity in the architecture works and the clothes. We can find clear traces of architecture in futuristic designs such as splicing acute angles, unconventional waved hems and towery collars, which greatly enrich our imagination. For example, in works of Hussein Chalayue, no usual seams are seen. Streamlined profiles shaped by three-dimensional cutting technology bring about artistic beauty at sight, creating soft "future architecture" in fashion designs.

设计漫谈

荀子说：学不可以已；青，取之于蓝，而青于蓝；冰，水为之，而寒于水。

设计无止境。源自于生活，而高于生活。设计是青，生活是蓝。然而，设计，改变生活。今天的成绩总是来自昨天的努力，或成功，或挫折。谁都会有一两个故事，或小，或大。

鞋柜的故事

曾记得，刚刚涉足业界时，依旧青涩，以为完美地憧憬和演绎生活，就可以将业主的雅居雅舍铸造成为一个完美的家居。然而，竟不曾注意，尺度完全合乎用度的鞋柜，居然因为分隔的间距过大，而使得本应够用的鞋柜空间变得局促起来。直到业主提出疑问，我才注意到。一个细节的疏忽，或许不是我的责任，然而，我本可以做得更好。

于是，这成为我的一个隐痛，更成为我的动力。多年以后，我和业主成为很好的朋友和合作伙伴。虽然每每提起这件事，总是一笑而过。可是，内心依旧受着鞭策。设计源于生活，生活由无数细节组成。于是，设计也要关注细节，才能改变生活。

撒哈拉沙漠的颜色

提起撒哈拉沙漠，几乎是家喻户晓、妇孺皆知的。然而，我竟然不记得，一望无际的撒哈拉，它是金色的！不是我不记得，只是，我一时间困惑在设计常态的怪圈里，我没有看到的是设计之外广阔的生活和天空。

酒店的颜色，在我的常识里，深啡色，无疑是富丽、豪华和高贵的。但是，当我将我的一稿方案给甲方负责人看的时候，他说："在中国，在哪里，这个方案都是一个很成功，很优秀的设计。"然而，他说"然而"，于是在 11 层的楼上，他拉开窗帘。一瞬间，我如醍醐灌顶。

是啊，我的设计在哪里都是优秀的，富丽贵气的颜色也没有错。然而，这是在撒哈拉的映照下啊！这是在中东啊，设计 AMLAK 酒店的时候，我考虑了地域特色，考虑了他们曾是英殖民地国家，崇尚的是欧式的美和审美。然而，我却忽略了一点，地域特色不仅是建筑的风格，还有颜色！

窗外是一望无际的金色，这种颜色的建筑都隐没了，远处的白色反而显得特别醒目。是环境影响了常态下的判断，在国内清一色的办公建筑的白色，放在这里反而变得出跳、出色、出彩！

最后，AMLAK 的曲折演变成为白色铝幕、黑色玻璃演变为撒哈拉金色中的一抹亮色！

甲方很满意，我也更加意识到设计根植于生活，而环境给了审美潜移默化的影响。

说到审美，说到设计，美是相似的，设计是相通的。因为，它们都是生活的触角。建筑是庇护，衣服也是庇护，一个是整体的，一个是个体的。时尚是相通的，就像未来的建筑和时装。柔软、流畅、起伏的弧线创造的是梦幻般的如同时装带给人的时尚和舒适的感觉。建筑师给建筑表皮亦如时装设计师般精巧、流畅、细致、创意的剪裁。未来建筑，是穿着时装的建筑。

相通的是未来的建筑和时装，然而，这是开始，不会结束……

因为，建筑设计、时装设计，一切的设计都是设计，都是人类审美的取向。同源于生活的它们，怎么会不相通、不相融？"大同"，或许是未来一切设计的发展趋势。

我们会在"大同"的设计世界里关注细节，关注生活，日臻完美，日趋完善，然后在生活中游刃有余地设计。而设计最终，将改变原来的生活状态，创造出一种熠熠生辉的生活。

因为，我们是设计师，我们从生活中来，我们在潮流的尖端，我们将改变生活，自己的，别人的，世界的！

Thoughts on Designing

Xunzi said:
Learning should never stop.
Blue dye comes from the indigo plant,
yet it is bluer than the indigo.
Ice is made from water,
yet it is colder than water.

Designing should never stop, either. It comes from the daily life but also transcends that. Design is the "blue", and the daily life is the "indigo plant". And design changes life.

The achievements we have made today all derived from the efforts we paid yesterday. The efforts, no matter end up to be successes or failures, are all blessings. Everyone has stories that influence their lives, no matter they are trivial or important.

Story of the Shoe Cabinet

I remember in the early years of my career life, I was not thoughtful enough. I assumed I could design a perfect house by holding expectations toward life and trying to interpret it. But I neglected the pragmatic side. The shoe cabinet that I designed, which should be large enough for use, became too small to use because of my improper design of the splitting space. I did not realize it until the client raised this issue. Maybe I didn't need to take the blame, but I failed to pay attention to the details. I could have done better.

This story has ever since been a thorn in my mind, and also a motivation for me to make progress. Years later, my client and I become good friends and trusty partners. Although we often laugh it off when this episode is brought up, I indeed learn a lot from it. Design comes from life, which is made up of numerous details. And we designers need to pay attention to the details if we are to change life.

Color of the Sahara

Everyone knows the Sahara. I do, too. But I forgot one thing: the Sahara is golden! To be honest, it was not that I forgot it, but I neglected it. I was trapped in the vicious circle of the design routine and failed to see the rich life out there under the sky.

I was trained to assume that the dark brown denoted nobility and luxury, but when I presented my first draft to the person in charge, he said, "This design would be an excellent one in any place across the China. However,…" He paused and drew back the curtain. We were on the 11th floor and I suddenly realized his point.

Yes, my design may be great everywhere and there was nothing wrong to choose the dark-brown color. But here we were in the Sahara, in the Middle East. When designing the AMLAK hotel, I paid respect to the local element and took into account their preference for the European style for they were once a British colony. But I neglected one thing: the local element is not only the architecture style, but also its color!

The golden desert stretched to as far as eyes could reach. Architectures in gold seemed to be disappearing, whereas the distant white stood out. The environment forced us to abandon the normal judgments. White, which was the main color for most domestic office buildings, became distinct, brilliant and innovative!

In the end, AMLAK turned out to be a striking artwork with white aluminum curtain walls and black glasses in the golden Sahara.

The client was very pleased and I also realized that the design had to be from life and that the environment would change the aesthetic perceptions of people in a slow but firm way.

When it comes to aesthetics, all forms of beauty are actually similar; and when it comes to design, all fields of design are interconnected, for they are both ways to reflect on life. The architecture is a shelter, and so do the clothes, except the former is designed for the whole, the latter, the individual. Fashion industry is interconnected, like the way future architecture does with fashionable clothes. Soft, smooth, undulating architecture arcs bring about a sense of fashion and comfort to people just as dreamlike fashion clothes do. The architect polishes the architectural skin in a delicate, smooth, meticulous and creative way like a fashion designer. The future architecture will be a fashion model.

Here we discuss the connectivity between future architecture and fashionable clothes and it is a starting point instead of a finishing line…

Architectural design and fashion design are both design works that reflect the aesthetic orientation of human beings. Originated from the life, they must be disconnected. "Datong", the traditional Chinese philosophical concept denoting the great harmony, may be the trend for all future designs.

In the "Datong" world of design, we pay respect to the details and the life, and keep improving, aspiring to design with ease in the life. And design will ultimately change the original form of life, making it shining and glorious.

For we are designers, and we come from the life and will lead the fashion. We possess the power to change the life, not only our own, but the life of others and the world.

设计圈的学习之道
— 关于读书和旅行

关于大师：

首先，我认为任何设计师，任何设计作品，无论大师与否，只要能引发对象心灵共鸣，激发起某些内在情绪的释放，正面的或者负面的，能够启发思考，那都有可能对我的设计创作产生影响。从学生开始学习到目前真正从事建筑设计这段时间，有三位大师的设计思想是融入了我的血液的。

柯布西耶在 20 世纪 50 年代后有很多优秀的设计作品，诸如朗香教堂等一系列表现主义的著作。对我个人而言，影响更为深刻的是这位大师 20 世纪 50 年代以前所提出的理论。他否定了古典主义和折衷主义，强调了设计上的功能化，与时俱进地强调了那个时代的机械美，提出了"住宅是居住的机器"的概念。他的作品永远在光影中追求，他的"新建筑五点"在一个世纪以后仍然为设计师们广泛使用，并且极有可能深远地影响下一个一百年、两百年。因此，柯布西耶这位大师，不仅仅是对我影响深远，在现代建筑教育体系下成长起来的设计师，绝大部分或多或少都接受过他思想的指导。

矶崎新运用的手法主义是真正意义上的表现主义。他给我的设计启发在于革命。摆脱了建筑传统美学一直在追求的"和谐，均衡，统一"，强调了"分散化，不和谐性、支架的间离"，他站在相反的角度去理解建筑，用黑暗挤压出空间，他认为城市是一片断壁残垣，他的看似怪物的建筑能营造出悲怆的气氛，他的作品用虚幻的空间体验去引发人类的自省。从这位大师身上能够看到一个建筑师的探险精神，以及对社会责任的诠释。

路易斯·康这位大器晚成的建筑师，他是一个诗人。诗人的创作不见得总是狂放不羁的，浪漫也不一定是情绪化的、粗糙的。观察他的作品，严谨的秩序表达出浪漫，质朴的气质透露出典雅，用最简单的几何形态追求着永恒，浪漫地展现出建筑的力量。值得一提的还有他设计创作理论与建筑作品的高度统一，这一点在当下的建筑师创作中是极其难得的。

关于旅行：

作为建筑师，必须经历很多次"出走"。有时是为了项目，有时是为了家人，有时是为了心灵自由。今年的夏末，为了项目设计，我们需要去了解傣族文化。这个旅程非常有意思。我们没有选择已经开发的旅游区作为目的地，原因很简单，旅游区为了迎合游客的需求，所展现出来的仅仅是最表层的符号，真正的民族精髓文化已被媚态掩盖。我们此次选择了云南和老挝边境一个步行3个小时才能到达的山寨。和世辈生活在这里的村民吃在一起，住在一起，我们和老人交谈了解他们的故事、他们的神话，我们测量并绘制当地没有被外界思想改造过的"原版建筑"。设计师从气候环境、生活习惯、社会礼教、建筑文化整体去尊重、理解并贴近一种文化，带着这种诚意，以这种文化为基础做出的创作一定也会受到人们的尊重。

第二个对我来说印象很深的地方是日本。我所在的城市是南京，很多朋友包括我个人在谈及日本一直持有有意或者无意回避的态度。如果要选择出国，一般情况下日本也不会成为选择，今年是我在日本的第一次体验。不做评价，说几个让我有触动的"小"事儿吧。初到日本，第一直观印象是那是一个拼装起来的城市，让我很感叹的是他们的工业化精细度的水准。我观察过他们的施工现场，首先围护隔离是非常严密的，在切割工作进行的同时雾化装置也在起作用，第一时间抑制粉尘扩散。与中国高调的大规模建造相比，日本的建造显得更谦虚内敛，尽可能地减少对工地周边的影响。这种谨慎谦虚的态度也体现在城市建设上，在处理新老建筑的关系时，也表达出以往现存的充分尊重，用技术手段去协调共存。当新建立交桥穿过建筑时，原有建筑仍然能保持正常使用，这种对资源的珍惜对技术的追求的态度也让我有所反省。一天清晨，我看到不远处马路边有一位老人步履缓慢，时而蹲下时而站起，以为他需要帮助，就留意观察了一下，走近了才发现，他是在拾路边的烟头。和他的聊天让我感觉对环境的保护深深地扎根在日本国民的灵魂深处，我非常感动。

关于客户：

"客户群体审美素质有待提高"，这个评论本身的提出，我个人觉得有一些片面，有失公允。在我所接触的客户中，大部分客户对美是有追求和认知的。现在网络书籍等信息非常普遍，审美这种是普遍性的社会活动，而不仅仅是受过专业系统训练的设计师的专利。当然，客户的审美水准和设计师存在差距，这是普遍存在的现状，但是在实际项目操作中，我认为完全有引导和协调的空间。同时，客户相对于专业的设计人，也有他们的优势。一个投资人要进行一个业态投资，在进入详细设计，落实到笔之前，已经做了大量的相关调研工作。当这些因素都统筹考虑过，达到胸有成竹的程度，才会对预想进行实施。其实，在这个过程中，他们已经形成了对此项目的初步预想。设计师站在纯设计的角度和客户沟通，也必须具备能够迅速与客户已有认知找到契合点的能力。相比与纯粹艺术空间的创作作为主要导向的设计师，很多时候，业主提出的对空间多种可能性的复合经营以及风险意识可能会要求设计师具有更发散的思维，激发出更多创造的可能性，对设计师的设计能力的提高也有帮助。

关于教育：

"供需不对口"这个问题的主要起源并不在学校。前几十年中国飞速地建造，使得我们必须快速提供大量的建筑产品。为了提高效益，减短建筑建造周期，我们被迫将设计这样一个整体创造的活动切分成很多工种，于是我们的学校教育也相应将建筑一门学科切分成很多专业去适应节奏。当狂热过去后，冷静思考，虽然我们的建筑产品多，但精品多么？建筑从外而内的完整度高么？受众使用的同时建筑的空间体验流畅么？建筑师懂城市么？室内建筑师懂空间么？设计是相通的，建筑学、室内环境艺术、园林艺术这些其实就是一门学科，都是处理空间与人的关系，无外乎大大小小罢了。很多人说"室内无理论"，我认为室内空间设计师要多去学习一些建筑的读物，对建筑学、园林、城市规划的学习会帮助我们了解整个建筑发展的历史，了解建筑空间的理论也会为室内空间的创造提供很多有意义的指导。无论专业怎么划分，学科怎么安排，高校教育中建筑理想的树立和学习方法的养成应该始终是排在第一位的。

关于读物：

我本人读书比较杂，什么类型的读物都愿意去学习。

叔本华是我很喜爱的哲学家，他的著作很多朋友早在中学的时期都已经阅读过，我现在也还会经常翻阅。近期在复习《叔本华思想随笔》。叔本华用朴素的语言，从哲学、伦理学、性学、美学、教育学、玄学、宗教等各个方面阐述了哲学思想。阅读他的作品让人懂得思想的力量，产生思想支撑世界的信仰。

《长物志》介绍的是明代的家具以及陈设，是一种独特的东方的生活观念。书里讲述的都是些琐杂碎细之物，不挡寒，不疗饥。但是在这些"宝贵的多余之物"上，能看出一个人有没有韵、才、情。没有韵、才、情的人，不能驭物，格调自然也就不同。文震亨此书的刊刻，既是对"富贵家儿、庸奴、钝汉"的当头棒喝，也堪称规范雅士格调的"法律指归"，所以沈春泽赞叹："诚宇内一快书，而吾党一快事矣！"

《蒋勋说宋词》的作者蒋先生首先是位美学家。所以读这本书觉得写得很美。本身也很喜欢李后主的词《浪淘沙》和《乌夜啼》。经蒋勋一讲，"梦里不知身是客"一下子就沉重了。相比前一部的争议，蒋先生的《中国美术史》非常通俗，不论老少都不会觉得晦涩难懂，从美学的角度分析艺术，从人性出发联系古今，以生动的文笔，按朝代次第勾勒出中国美术的脉络，传达出中国人对美的独特感受。

Ways of Study in Design Community
—
About Reading and Travelling

About master:

First, I believe that any designer, any design work, whether master or not, can influence my design and creation, as long as they can have a resonance for me, stimulate the release of some internal emotions, whether positive or negative, and inspire thinking. From the beginning of my student to the time when I really started my career, I have been influenced by the design philosophy of three masters.

Corbusier had a lot of excellent design works in the 1950s, such as a series of works of expressionism like the Chapelle de Ronchamp. However, the theory proposed by the master before the 1950s had a more profound impact on me. He criticized classicism and eclecticism, emphasizing the functionalization of design. As someone who kept pace with the times, he proposed the concept that a house was a machine for living in, highlighting the beauty of machine in that era. He always pursued light and shadow. The "Five Points of Architecture" proposed by him are still widely used by designers after a century, and are likely to have a far-reaching impact in the next hundred or two hundred years. Therefore, Corbusier is not only a mentor to me, but also designers who grow in the modern architectural education system. In fact, most designers have more or less received the guidance of his thoughts.

The design style of Arata Isozaki is real expressionism. The inspiration he gave me is revolution. Breaking away from the "harmony, balance and unity" that traditional architectural aesthetics have always been pursuing, he emphasized "decentralization, disharmony and the separation of scaffolds". Believing that cities were ruins, he stood in the opposite direction to understand architecture and used darkness to squeeze out space. The buildings designed by him look like monsters, but can create a melancholy atmosphere. His works spark people's self-examination with illusory space experience. From this master, one can see the spirit of adventure and the performance of social responsibility of an architect.

Louis Kahn, a late bloomer in the architectural community, is a poet. The creation of a poet is not always unrestrained, and his romance is not always emotional or rough. Observing his works, you will find romance contained in the rigid order and elegance hidden in the plain style. He pursued eternity with the simplest geometry and demonstrated the power of architecture in a romantic way. What is worth mentioning is a high degree of unity between his design philosophy and architectural works, which is quite rare in the creation of contemporary architects.

About travelling:

An architect must "run away" for many times, sometimes for a project, sometimes for family, and sometimes for the freedom of soul. At the end of this summer, we were required to understand the culture of the Dai nationality for designing a project. The journey was very interesting. We did not visit any tourist area that had been developed out of a simple reason: to cater for the needs of tourists, tourist areas only exhibit the most superficial symbols, while the real quintessence of ethnic minority culture has been covered by subservience. This time, we chose a village on the border between Yunan Province and Laos which cannot be reached without three hours' walk. We ate and lived together with villagers who had been living there for generations, talked with the elderly to understand their stories and myths, and measured and drew "original buildings" which had not been transformed by thoughts from the outside world. Designers should respect, understand and maintain close contact with the culture from its climate, customs, ethics and rites, as well as architecture. With such sincerity, their creations will certainly win people's respect.

The second place that leaves me a deep impression is Japan. I am from Nanjing. Many of my friends, including myself, always refrain from talking about Japan consciously or unconsciously. If I have to go abroad, usually I will not take Japan as an option. Without comment, let me say a few "small" things that touched me. When I arrived in Japan, my first impression was that it was a city pieced together by fragments. I marveled at how meticulous their industrialization was. When I was observing their construction sites, I found that the containment was quite strict. The atomization device was working while the cutting work was carried out, so that the dust could be controlled the first time it was produced. Compared with China's high-profile and large-scale construction, Japan's construction, which tries to minimize its impact on the surrounding area, seems more modest and restrained. This cautious and modest attitude is also reflected in city construction. In dealing with the relationship between old and new buildings, the Japanese also show full respect to the existing buildings and try to achieve their harmonious coexistence with technical means. When an overpass is built through a building, the building could still maintain its normal use. This attitude of cherishing resources and pursuing technology also makes me reflect. One morning, I saw an elderly man walking slowly not far on the road, sometimes squatting down, sometimes standing up. I thought he needed help, so I watched for a while. When I approached, I found that he was picking up cigarettes at the roadside. Through chatting with him, I felt that the awareness of environment protection was deeply rooted in Japanese minds, and I was deeply moved.

About clients:

Personally, the opinion that "clients' aesthetic judgments need to be improved" is a little one-sided. Most of my clients pursue and can appreciate beauty. As people now have access to various information like online books, appreciating beauty is no longer the exclusive right of designers who have received professional and systematic training, but has become an universal social activity. Of course, there is a gap between the aesthetic judgment levels of clients and designers. This is a common situation. However, I believe that there is room for guidance and coordination in actual projects. Meanwhile, clients have their own advantages over professional designers. An investor must have done a lot of investigation before investing in something. They will not put their plans into action until they have considered all concerned facts. Designers who communicate with clients from a pure design perspective must also have the ability to quickly find the common ground they shared with the clients. Compared with those who are purely oriented toward artistic creation, designers whose clients have risk awareness and ask for multiple uses of space may require more divergent thinking, inspire more creativity, and improve the designers' capability.

About education:

The imbalance of supply and demand in education is not derived from school. China's rapid construction in the past decades demands us to provide a large quantity of construction products very quickly. To improve efficiency and shorten the construction cycle, we have to divide design, an integrated activity, into many types of work. Therefore, architecture in education has to be divided into various disciplines to adapt to the society. After the fever, let us think calmly. We did produce many construction products, but are there many quality products? How integrated the buildings are from outside to inside? Do people have a smooth experience in the buildings? Do architects understand the city? Do interior architects understand space? Everything in design is interconnected. Architecture, the artistic design of the indoor environment and landscape design are in fact one discipline, because they are all about dealing with the relationship between space and humans, except that the scale is different. Many people argue that "there is no theory about interior design", but I believe it is necessary for interior designers to read some books about architecture. The study of landscape and city planning can help us understand the history of architecture, and understanding the theory about space can provide guidance for our creation. No matter how disciplines are divided, the education on architectural ideal and learning methods should always be taken as a priority in colleges and universities.

About readings:
I read all kinds of books, and I am willing to read.

Schopenhauer is a philosopher that I like a lot. Many people have read his books as early as in middle school, and I still read his books very often. Recently, I have been reviewing *Arthur Schopenhauer*, which explains philosophical thinking from the aspects of philosophy, ethics, sexology, esthetic, pedagogy, metaphysics, religion, and so on. Reading his works helps one understand the power of thought and makes one believe that the world is supported by thought.

Treatise on Superfluous Things introduces the furniture and furnishings in the Ming Dynasty which embodies a unique oriental concept of life. Everything in the book is trivial thing that can not keep one warm or allay one's hunger. However, from these "valuable unnecessary things", one can tell if a person has charm, talent or love. Those who without charm, talent or love can not rein things, therefore have a different style. This book is not only a sharp warning to the offspring of wealthy families and ignorant people, but also a code of conduct on style for refined scholars. Therefore, a scholar in the Ming Dynasty praised it as "a great book in the universe and a happy event for people".

Jiang Xun's Comments on Song Poems is written by Jiang Xun, an esthetician. Therefore, when I was reading the book, I found its style of writing is quite elegant. The poems written by Li Yu, the last ruler of the Southern Tang state and a famous poet whose works I admire a lot, become more melancholy after Jiang's explanation. Compared with this controversial book, *History of Chinese Art* is more popular and easier to understand. It analyzes art and history from the perspective of aesthetics and human nature, drawing an outline of Chinese art divided according to dynasties and conveying the unique feeling about beauty of the Chinese people with a vivid writing style.

现代装饰
2014 年 5 月刊·对话
来自上海读者的提问

我理解中的现代与传统结合的建筑艺术，在一定意义上并不应该仅仅属于一种"中间状态"的现代建筑类型，或是被简单地视为"现代风格略带中国色彩"。中式建筑，即便是现代建筑，它亦必须传承"中国固有式"传统思想，符合中式审美及思维情趣。

是不是传统的一切都归于大屋顶、小亭子、假山池塘、半亩园呢？从创作者的角度，个性不同所产生的具体表现肯定也存在差异，"活"是中国建筑处理手法的一个特别明显的特征。中式建筑基本不会出现充满暴力的墙体，建筑气质不显热烈亦不冷漠，与西方建筑建筑空间包围建筑不同，他更强调的是建筑引领出整体空间。西方思想突出的是建筑的本体，而中式建筑则往往更享受禅宗意味的"空"。当作者对中式建筑有充分的理解，便会觉知，形式对于空间只是表皮，创作者传递给观者的应该是一种意蕴，想象之中，描述之外的体验。

在旧建筑改造项目开展的第一天起到竣工交付，每一天都有可能出现各种各样的困难挑战。最基本的一点，旧的躯壳要载入新的灵魂，要让旧的建筑能够承载新的功能使用，这就需要对旧建筑进行空间重塑，重塑的目的在于优化而不是破坏。介于城市对于旧建筑的情感因素，应该尽可能地维持建筑本身的艺术气质，维护其本身的历史延续性，在对原有文化空间和历史认同的前提下，合理利用先进技术和材料，再构我们现代生存空间的框架。举个例子，要适应现代生活需要，室内各种管道的设置便是不能回避的一个大问题，这就需要根据具体情况制定具体的方案，或藏或露，或整或零，都考验着设计师对于整体项目的把握。

May,2014 Dialogue Column
Questions from The Readers
from Shanghai

In my understanding, the architecture style that combines modern and tradition does not necessarily belong to an intermediate state of the modern design, nor should it be considered simply as "modernistic with a slight Chinese tint". Chinese architecture, even a modern one, should inherit the "innate Chinese" traditional philosophy and conform to the Chinese aesthetics and thinking pattern.

Does all traditional elements come down to big roofs, delicate pavilions, rockeries and ponds, and half-acre gardens? From a creator's perspective, the specific presentation of individual architecture certainly varies. "Flexibility" is a distinct feature of Chinese architecture design technique. Violent wall forms are nowhere to be seen and the temper of the architecture appears to be neither enthusiastic nor aloof. Compared with the Western architecture space where the space encompasses the architecture, the Chinese architecture emphasizes more the function of the architecture to lead the whole space. Thus in Western philosophy, the architecture itself is to be emphasized, while the Chinese architecture seems more at ease to live with the philosophical state of "emptiness". If the designer has a deep understanding of the Chinese architecture, he will come to understand that the external form is only a cover for the space, which is second to the metal experience that the designer can deliver to the audience, which is beyond description and imagination.

When the old building is reconstructed, from the very first day of its operation till the end, challenges and difficulties may wait at every corner. And the basic principle is that old form needs to be instilled with new spirit to bear new functions of the architecture, which involves the reshaping of the architecture space. The aim of reshaping is to optimize the existing space instead of spoiling it. Given the emotional attachment the city holds toward the old architecture, designers should endeavor to preserve the inherited artistic feature of the architecture, maintaining its historical continuance, and reconstruct the framework of the modern living space under the premise of historical identity with due use of advanced technology and materials. For example, the organization of interior pipelines is a major problem for modern living needs. It requires designers to choose appropriate plans for specific situations. To hide them or to expose them, to put them together or to break them down ? The final decision lies in the designers' overall design of the whole project.

聚焦江南
回归东方精神本源

采访手记　许晓东
媒　　体　设计家

设计界每一股潮流的兴起，都与社会生活方式、审美思潮趋向息息相关。重新定义中国人的生活方式，这一话题在当下较为引人关注。所谓的重新定义，其实是对中国社会文脉断裂、财富爆发以来种种无所适从的乱象的梳理和反正，是对中国文化精髓与当代生活结合的探索研究，颇有"与古为新"的含义。

青年设计师潘冉，近年来创作的《竹里馆》《小东园》《来院》等代表性作品，以当代设计手法表现具备中国精神气质的空间，将新与旧、轻与重、形与神、雅与庸进行了很好地结合。潘冉生长于皖江中南部的鱼米之乡，此山此水此食，天人合一的生存智慧，从身心的本源处滋养着他的成长。传统文化的浸润细腻无声，却深入骨髓，多年后终现端倪。大学时醉心摇滚，在音乐领域玩得风生水起，毕业时毅然回到本专业，进入南京的设计院从学徒做起。潘冉的设计历程和同时代的很多设计师一样，在城市急速膨胀的背景下，经历了大体量、装饰性、快速建造的商业性项目的浪潮冲击，同时也经历了一个不设方向如饥似渴的学习过程。但是多而快的商业性项目并没有让他感受到快乐和满足，反而让他内心充满了焦虑不安。通过系统地梳理东西方设计脉络，多方位、多领域地潜心修习，潘冉带领着他的团队聚焦东方文化，研究属于中国人自己的生活方式和美学观念，血脉与文脉的契合使得这一过程舒服而顺畅。2013年以来，潘冉经历了从对西方方法论的尊崇到对中国精神认识的回归，从不知徽派为何物，到骨子里透出古典的气韵，从西方的摇滚热爱到中国传统琴学与书法的精通，

2015 年他们把范围缩小，聚焦"以江南文人的精致生活态度解读空间美学与哲思"——这是内在心灵的需求促成的改变。当设计从宏大叙事变为深入叙事，更多地关注、解读精神层面，宁少求精，当设计做减法，有一种快乐便慢慢地浮现。

潘冉希望中国的空间设计来一场革命，也希望扫一扫美盲，避免民众因对生活方式和美学不自信而耽于盲目复制舶来文化。设计，不是进行所谓的"装修"，即装点修饰；而是回归到《园冶》里所说的"装折"，即从室内到室外的转折去重新认识一个空间。他相信这样带来的是完全不一样的状态。

请谈谈您的成长和学习经历。您从安徽建筑工业学院（现更名为安徽建筑大学）艺术设计系毕业后为何会到南京发展？

我从母校毕业的时候，内心非常迷茫。大学四年我没怎么认真上课，在外面租了个房子搞摇滚。虽然当时音乐上发展得挺好的，但我觉得搞音乐太漂泊，还是想回到专业上来。快毕业的时候，我们家在南京的亲戚答应帮我找一份工作。印象中，南京是一个很有底蕴的地方。我想那就去吧，因为一份工作。

您出生、成长、学成于安徽，在这二十几年里，徽派文化对您美学基础的奠定以及价值观的形成有哪些影响？

客观地说，在我产生独立认知之前，徽派文化对我的影响是不可见的，虽然它每天都在我的身边。多年后，我开始接触中国的传统建筑，然后再回过头来看徽派文化，才发现原来徽派文化是深入我骨髓的。儿时，身边的好山好水、好味深长的鱼米之食，这些人文的东西，是从生活中自然透露出来的气息，是一种生活习惯累积而成的美学观念。做设计很多年以后，我开始越来越喜爱中国传统的东西，我开始知道当年时时在我身边的这些东西是如此宝贵。现在，我会把当年自己的切身感受联系在一起去看待徽派文化以及我脑海中的美学观念。没有认知是很可怕的。在建立世界观之前，你会觉得所有的发生都是理所当然的。但是当你建立认识事物的判别标准以后，你觉得很多东西来之不易，而且它的发生一定是有迹可循的。

能谈谈早期的设计吗？

早期，我经历了大部分设计师都会经过的过程。当学徒，在图版上画图，学习软件制作，跑工地，慢慢会接触一些小型项目。你需要迅速地在团队中做到技能最好，慢慢老板会赏识你，觉得你能干，给你更多的项目。当你想着如何做创意的时候，你就进入第二个阶段了。这个时期你会获得更多的机遇？如何把握每一次机遇？把图纸画好、把工作做细，或者是创业！在这个过程中，你可能会成长为一名设计师。当时在江苏省建筑设计研究院工作，随着工作越来越得心应手，我开始带领团队做投标，做面积比较大的大型项目，同时接触社会中不同的人。2007年，我开始和几个同事组建工作室，那时候还没有注册公司。虽然刚刚起步，但陆续有人来找我们做项目，而且都是面积偏大的，广场、综合体，也有建设集团委托我们出创意。2008年，我们正式注册成立事务所。

2007年，您曾参与外交部援建项目，往返迪拜、阿布扎比、开罗、喀土穆，这段经历给您带来了哪些影响？

对，那时候正好有一个建设集团帮助中非地区援建，建外交官俱乐部，还有一些小型精品酒店。这段经历挺精彩的，我参与援建项目近两年，在这个过程中接触到很多有意思的人，这有什么好处呢？打开视野。原本可能固化的模式会随着工作和面对的人的转变被很快打破。我当时写过一篇文章《设计漫谈》，里面有写到"撒哈拉沙漠是黄色的，它更需要白色的建筑"。

应该是在此之前您已经有了不错的作品，业主才会慕名而来吧？

当时已经有了一些作品，这些作品现在看来非常商业，但当时大家觉得好。我2009年回国，2010年团队开始区别对待大型项目和精品项目，越来越少做体量大、很商业的快速项目。我们更多地做中小型的、精品的、要求很高的、能够有独立思考的项目。

名谷设计经历了哪些发展阶段？

名谷设计从成立到现在经历了两个阶段，有可能即将进入第三个阶段，前两个阶段我们从宏大叙事转到深入叙事。宏大叙事，可能更多的是一门生意，既是设计又是生意。深入叙事，是真正扎入设计工作，它的思考是深入的，是有深度的设计。这几年我们进入深入叙事的阶段，接下来，我们希望能整理近十年的项目，有一些思考，并形成文字，推出一本作品集。

从商业转变到更多精神、文化层面的追求，势必要做一些取舍，取舍背后的内因是什么？

当时明明做了那么多项目，内心却没有觉得很充实，时常感到焦虑与困惑，我想一定是哪里出了问题。可能是我们走得太快或要得太多。刚开始转变是要付出代价的，这种代价来自方方面面。经济上的最明显，其次是精力，大型项目回报更多，做精品项目，你首先需要投入好几倍的精力揣摩、推敲，才能落地，回报却不显著。我们经历了一到两年这样的蜕变。蜕变的过程其实挺痛苦的，但是内心却越来越充实——这才是做设计。后来我们坚持以这样的工作方式去完成每一个项目，也要求每一个项目都是精品，能够成为作品。设计，就是做减法。刚开始做了很多，后来越来越少，越来越简单，会越来越直接地表达自己的想法。做减法以后，有一种快乐会慢慢浮现出来。

能总结一下您的设计哲学吗？您的设计哲学经历了怎样的形成过程？

在没有建立自己的价值观、设计观之前，学习是不设方向的，东方的、西方的、本土的，甚至少数民族的都去学习，可以说是如饥似渴。后来去了世界各地游学，当你亲临先辈大师的作品现场去感受那个空间透出的气息，慢慢地你会感受到有一种气息与你一见如故，这就是东方的能量场悄然而至了。我们开始把学习方向聚焦于东方，研究东方的生活习惯，觉得很舒服、很顺。因为你骨子里面流淌的就是东方的血液，包括你的生活方式和美学观念。你很容易找到契合点。我们开始研读建造和生活美学类经典，比如《营造法式》《园冶》《长物志》等，这些都是宝贵财富。可惜当下的设计教育不重视这些，我们在学校学的都是当年包豪斯模式下的产物，是西学东渐的产物。

2013 年开始，我们团队的设计价值观就确定为东方人文空间。我认为最精彩的东西就是传统精神与当代践行，两者结合起来才能爆发出一种能量。2016 年我们继续深入，缩小范围，聚焦江南，提出"以江南文人的精致生活态度解读空间美学与哲思"。西方人对方法论的更新非常迅速，但在精神认识上我们需要回归。西方的方法论、建造技术以及对空间结构的理解都可以拿来用，但不能失掉东方精神，如果能中西合璧就非常好。后来我发现中国有很多世界级的建筑师，虽然在西方留学，但回来后做的东西依然有东方的能量场。比如说冯纪忠先生，他设计的方塔园让人百看不厌。那才是真正的中国建筑精神，宋的气韵从石缝里透出来，非常强大，在我看来，目前中国这么多建筑没有一个可以超过他的"何陋轩"。去年上海有个展叫《沉潜的现代》，展出了冯纪忠先生和王大闳先生的作品，他们用建筑来思考时代的意义！

现代主义的建筑非常让人害怕，我觉得非常危险，像北京或某些发展特别快的城市，包括一些很有历史价值的城市，几乎成了建筑的实验场。2014 年，我在上海东华大学做过一场演讲，我说中国当代建筑的评判在某个时间内是以刷新标高来决定的，而不是以美学或适合与否来衡量。再回到江南，以前的人，达官贵人或普通老百姓，在某一个时代，生活都很讲究，有仪式感，每个生活细节，书桌放在书房的哪个位置，窗底下是鱼池还是种竹子或芭蕉。讲究的人对自己有要求，对别人是礼貌的。然而我发现当代很多成功人士，虽然非常有钱但是享受不到真正美好的东西，这是一件非常可悲的事情。这个设计也是有责任的。文盲扫得差不多了，该扫美盲。大家对美学、生活方式表现出一种不自信。现在的住宅设计似乎失去了自我，盲目地把西方的东西拿来，只是复制，美式、欧式、法式，这些风格需要一个大的客厅，有一个壁炉。但是，这样的房子，我们中国人住着舒服吗？这在外国设计师看来是非常可笑的，明明我们有那么强大的底蕴。模仿是可以的，但它只是一个阶段，需要快速通过，而不是停留在这里全面展开。有很多东西需要去沉淀，去和当代结合起来，开展当代的生活方式。在设计的当下阶段，如何找到自我，完成自我，就像心学大师王阳明说的完成自我的良知非常重要。孔子尚《周礼》，魏晋到了思想的高峰，南宋、北宋到达了艺术的高峰，明朝有文化复辟和心学，现在，我们应该先建立自己的价值观，从自己出发，设计出自己的东西！

现代化的公寓其实是二战后的设计师研发出来的一个满足居住功能的产品，它强调居住功能，但不负责情感问题。可是中国的园林大都是从美学和情感开始建构的，它负责沟通人与物的情感问题，负责沟通人与自然的关系。我们的设计不仅要功能化，也要美学化和情感化。我觉得只有把三者糅合在一起，住宅才是真正的心身归宿。

您怎么看传统文化和当代审美、生活方式的融合？

传统文化和当代审美具有共性，它们都是从生活中流淌出来的，都是经历了时间的筛选而留下的精华，如果你的生活方式是相近的或类似的，你一定会从中感觉到一种美好。现在的生活方式随着社会结构的改变而变化，但是我们的衣食住行是一代代人传下来的，它一定会有共性的东西，我们只有把传统的东西拿到当代来用，才可以让它发光发热，否则它可能只是一个摆设。传统文化最精髓的部分就在于给你精神上的指引，特别是当你不知道何去何从的时候。我们为什么要去接触好的艺术，读唐诗看宋画，《溪山行旅图》一个瀑布从山涧顺流而下，气韵便产生了，瀑布几经转折流入一片虚像之中，那是留白想象的地方，空间便产生了，虚像里面停留的各种想象正是中国精神的表达。为什么现在人们开始喜欢插花、喝茶？因为这些可以提高你的精神品格，精神品格提高了，人变得高雅了，才会去思考如何生活。

您如何理解先锋和传统之间的关系？

我们一边敬畏传统，一边尝试先锋，一边学习古人的精华，一边解读当代的行为方式，进化和退化都不可取，也都可取，最终的目的是为了造化，通过一前一后的关系，去形成一种化境，这个化境是属于我们自己的造化！

"先锋设计师"这个称号，是外界对您的定义还是您自己的定位？

几年前有一些媒体朋友、一些评论人是这样介绍的，说着玩的，在我们团队的介绍里会出现这个词。我看来不太准确，先锋从物理学上来说是不成立的，谁能超越时间呢？但是它是具象表达的一种方式。

您在南京从事设计行业已有十多年，创立名谷设计也有近十年，您怎么看南京的设计氛围？

南京的设计氛围和其他城市比，可能比较低调，没有特别频繁的交流。有一些小圈子经常会有小范围的讨论，但基本是局限于南京之内的声音。南京有好的媒体和设计师，《Id+c》杂志就是南京出的。去年，我组织过几次活动，邀请我的朋友来南京做演讲。我也带着他们去杭州、福州等城市对话交流。现在信息越来越开放，我相信以后的交流会越来越多。

南京作为六朝古都，又有十分浓厚的现代气息，那么在保护和发展历史街区的问题上，设计师们会遭遇哪些矛盾？如何取舍？

我从事这样的工作有好几年了，有一些心得，重点在于机制：保护多少？发展多少？我们是把古董留下来当展品，是把它重新利用起来，是造一个假的古董让人去回忆？这些都不是设计师个人能够完全决定的。打造一个历史街区，重点是有什么样的机制去让政府、专家、设计师们做出客观的判断。比如，古建专家说这个要保护，不能动，但开发商会说我要能够使用，设计师要做的则是平衡。保护要让它产生价值，也就是说历史建筑如何在当代使用？如何能用还不破坏它？这就是设计师该起的作用。设计师在下判断之前，必须懂历史，了解建筑的结构，哪个地方能动哪个地方不能动，更要结合当代高超的建筑技术，通过技术来达到诉求。但我们一定要尊重传统的建筑，不能破坏它。

<u>可以谈谈竹里馆的设计缘由、设计理念吗？您的设计灵感来自什么？</u>

前年，我在思考怎样用最廉价的物料、低廉的成本去打造空间，以民间土木的方式完成建构。刚好我有个朋友，一个女作家想做个宅子，当时在想能不能打破现有的住宅概念，不要客厅，进来就是大书房，里面摆满书，用一种物料解决所有问题，我开始对竹子感兴趣，最后实现了当初的想法。后来竹里馆的创办人黄伟先生找我做茶馆，他也提出以竹为主题，由此一拍即合，顺势而为！

<u>竹里馆设计的重点和难点分别是什么？</u>

竹里馆设计的重点是简单快速，把特别简单平凡的竹子，通过排列让空间产生秩序感，让一个很普通的空间产生一种有秩序的内在逻辑关系。如何能够获得弹性的光线是重点。难点是需要解决很多工法的问题。我们要求每一根竹子都可以拆卸，拆卸的点需要我们设计很多新的小节点。竹子不接触地面，通过一根钢支撑竹子。每一根竹子都可以轻松拿下来。竹与竹之间的交接关系、防虫防裂，这些东西之前我们都没有做过，通过这次实践，我们也探索性地做出尝试。

您设计的《竹里馆》《小东园》《来院》等代表性作品都是具有中国精神的，可以谈谈您如何把握并塑造一个空间的整体气质吗？

中国的东西，很多形态都是气韵的表达，并不具象。很多人认为中式可能是一个花格窗或者是雕梁画栋，这其实是一层皮，重点是表达这个空间的气韵从哪来？这是骨美。就像我们看宋徽宗赵佶《桃鸠图》，鸟看的那个方向是气韵流通的地方。这幅画真正的意境就在留白的地方，它可以让你展开想象。

可以具体谈谈"空间－陈设－灯光三位一体设计模式"吗？

空间、陈设、灯光三位一体，是我们2010年提出来的。经历过大量的案例经验以后，我们觉得任何一个空间的表达都应该完整地被体现。这种完整体现在各个维度上，比如说空间里面要有家具，家具需要有灯光去表现，眼睛看到的地方要有光。这三个单元缺一不可，必须体现一种完整性。

在设计中，您怎样将科技与艺术融入其中？

一个好的建筑应该是既好用又好看的。当年包豪斯说过一句话，建筑师有责任把民族手工艺者和艺术家从城市的边缘解放出来。这句话对我影响挺大的。所以我在项目中会首先解决各种功能问题，然后，我会邀请艺术家、手工艺者一起参与，让空间具备一种独特的气质或艺术性。

您目前的设计项目类型有哪些？是否有所侧重？

目前的项目类型主要是精品酒店、房产销售中心、样板单元，不会特别侧重什么类型，只要是有想法有追求的业主，我们都希望为他呈现不同类型的作品。

您怎样管理团队以及管控项目质量？

我们团队管理没有特别的条条框框，而是围绕项目展开，比如有一个新项目，我们会为它建立一个单独的团队，有效分工。在控制项目方面，我得出一个心得：唯细不破。把所有需要考虑的环节非常清晰地表现在图纸或设计文件上，通过有效的控制，基本都可以落地。

除了出书之外，您对名谷设计未来的发展规划是怎样的？

我希望把"深入叙事"走下去，把根扎得更深一些，更扎实一些。最后能够把"中国式的设计"这句话给说清楚，通过我们作品的不断落地、不断去尝试与推敲。

业余您喜爱做什么？您认为设计与生活的关系是怎样的？

业余，我爱和艺术家交流，喜欢看各种各样的画展，喜欢弹古琴，喜欢书法。在艺术中我会获得很多养分，这对设计也有帮助。我以为设计源于生活，也可以高于生活、引领生活。设计和生活的关系则一直在变化中。

其他还有什么您想表达的？

当下中国的建筑设计可能需要一场革命。我们对装修这个词的理解是装点修饰，还是像《园冶》里所说的"装折"，从室内到室外的一个空间转折去重新认识一个空间。我相信这带来的是完全不一样的状态。

2013 年	* 中国外交部援建项目设计师，往返于迪拜、阿布扎比、开罗、喀土穆
	* CIID 中国建筑学会室内设计分会授予中国50 位室内建筑师奖
	* 《室内设计与装修 id+c》杂志 2013 年度中国新生代设计师十人

2014 年	* CIID2014 设计师峰会上海站设计师公开课演讲嘉宾
	* 2014 中国（上海）国际建筑及室内设计节金外滩奖、最佳景观设计大奖
	* 第十二届（2014）现代装饰国际传媒奖年度休闲空间大奖、年度软装陈设空间大奖

2015 年	* CIID2015 设计师峰会太原站设计师公开课演讲嘉宾
	* APIDA 香港亚太区室内设计大赛酒店类EXCELLENCE
	* 第十三届（2015）现代装饰国际传媒奖年度公共空间大奖

2016 年	* 2016 日本 JCD Design Award 全球 100银奖
	* 2016《精品家居》杂志 BEST100
	* AWARD 中国最佳空间设计大奖光华龙腾奖——2016 中国装饰设计业十大
	* 杰出青年（国家级）
	* APIDA 香港亚太区室内设计大奖第十四届现代装饰国际传媒奖年度餐饮空间大奖

2017 年
- 德国 Red Dot Award 红点设计奖
- 德国 WAF 世界建筑节奖、INSIDE 世界室内设计奖
- 美国建筑设计奖 American Architecture Prize 年度设计公司
- 台湾 TID 室内设计大奖
- 香港 *Perspective* 杂志颁发的 40under40 亚洲室内设计卓越设计师奖
- 英国 Andrew Martin 国际室内设计师奖
- 意大利 A Design Award 设计奖建筑类金奖、室内设计类金奖
- 美国 IDA 国际设计赛事建筑类和室内设计类特别提名、银奖、铜奖
- 伦敦设计奖 London Design Award 金奖
- 英国著名室内设计杂志 *FX* 组织评选的 FX International Interior Design Awards 国际室内设计评选（Mixed-use development）组别的最终国际大奖得主，也是 2017 该奖唯一华人获奖者

2018 年
- 德国 IF Design Award 设计奖

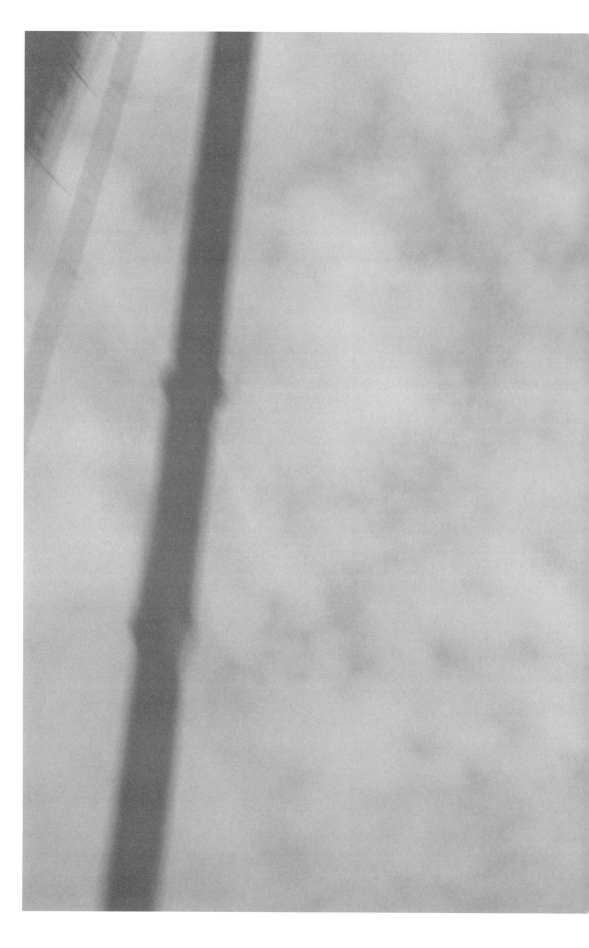

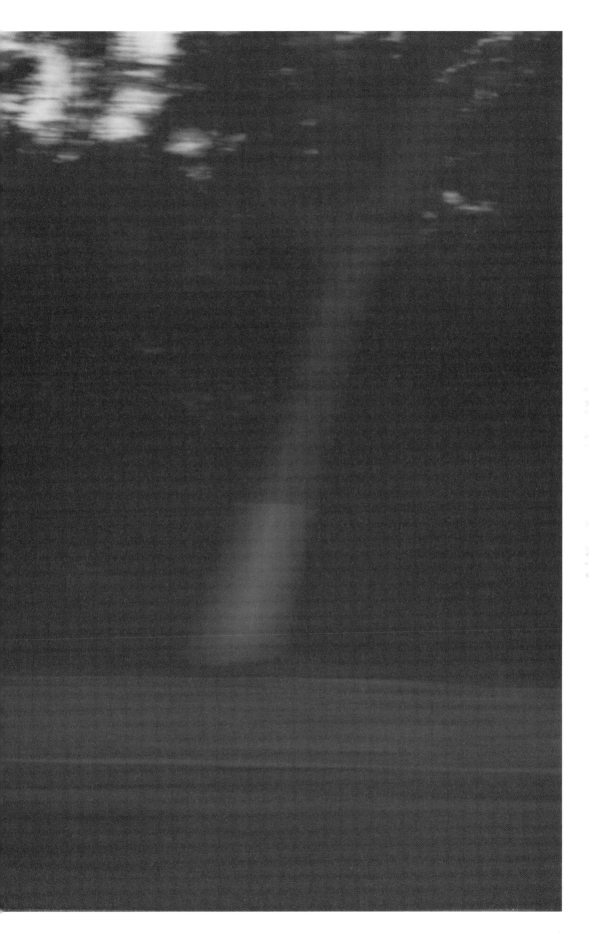